John Burgoyne

A Treatise on the blasting and quarrying of Stone

For building and other Purposes

John Burgoyne

A Treatise on the blasting and quarrying of Stone
For building and other Purposes

ISBN/EAN: 9783337106799

Printed in Europe, USA, Canada, Australia, Japan

Cover: Foto ©ninafisch / pixelio.de

More available books at **www.hansebooks.com**

A TREATISE ON THE

BLASTING AND QUARRYING OF STONE

FOR BUILDING AND OTHER PURPOSES

WITH THE
CONSTITUENTS AND ANALYSES OF GRANITE, SLATE LIMESTONE, AND SANDSTONE

TO WHICH IS ADDED
SOME REMARKS ON THE BLOWING UP OF BRIDGES

By Gen. SIR JOHN BURGOYNE, Bart., K.C.B.

A New Edition

LONDON
CROSBY LOCKWOOD AND SON
STATIONERS' HALL COURT, LUDGATE HILL
1895

ADVERTISEMENT.

THE importance of blasting, almost the only improvement which modern science has placed at the disposal of the quarryman to lessen his toil, has already exhausted several editions of this treatise; and the practical knowledge brought to bear on the subject by General Burgoyne, which it communicates, is so highly appreciated by all connected with stone and slate quarries, that a reissue of the treatise is again called for.

To the fifth edition an Appendix was added by Mr. Weale, containing some account of the failure of the attempts to blow up the unfinished bridge of Pesth by the Austrians in 1848; and of a like unsuccessful attempt by the same to blow up the bridge over the Ticino, near Magenta, in 1859. The first was communicated by the late Mr. William Tierney Clark, the engineer of the bridge; and the second is from the pen of General Burgoyne.

Mr. Weale also added the parliamentary speech of Captain Vernon when calling the attention of the House of Commons to the "Services of the Royal Engineers in the Crimea." That speech will ever be read with interest as a tribute justly paid to the skill and discernment of General Burgoyne during that campaign, more particularly as regards the siege of Sebastopol.

By way of Appendix Mr. Mallet has added a note upon "Mechanical Rock-boring and Cutting," as pursued at the Mont Cenis Tunnel, now in progress, and upon "Rock-hewing Machinery" employed upon various coal-seams in the north, and at the Holyhead Quarries, where the rock is quartzite.

To this edition a general Index is also now added.

January, 1868.

CONTENTS.

PART I.

QUARRYING AND BLASTING ROCKS.

	PAGE
INTRODUCTORY OBSERVATIONS	1
BORING THE HOLES	2

Results in Granite Quarries at Dalkey, 2.—Churn Jumpers, 3.—Boring with Augers, 4.—German Boring Apparatus, 5.—Position of the Holes, 5.—Line of Least Resistance, 5.—Charge of Powder required, 6.—Best Positions in which to operate on the Rock, 9.—Instance of the Force of Powder at Dunmore East, near Waterford, 14.—Table of Space occupied by given Quantities of Powder, 16

THE POWDER AND THE CHARGE 17

Strength of Powder used, 17.—Table of Comparative Results of different Qualities of Powder, 18.—The Eprouvette Gun, 19.—Table of Experiments on Powder by Bursting of Spherical Case Shells, 20.—Advantages of Using Superior Powder, 21.—Mixture of Sawdust with Powder, 22.—Leaving an Air-space adjoining the Charge, 23.—Mixing Quicklime with Powder, 24.—Mode of operating on the Hard Limestone of Gibraltar, 25.

LOADING 26

Ordinary Methods of Loading, 26.—Marked Copper Measures for Charges, and Copper Tubes and Funnels for Filling, 28.—Tamping-bar to be tipped with Brass, 29.

CONTENTS.

	PAGE
THE TRAIN AND FIRING	28

Safety Fuse, 30.—Account of its Use at the Kingstown Harbour Works, and Description of the Invention, 30.—Firing by a Voltaic Battery, 34.—Mode of Lighting a Train in Shafts, 35.—Means for preventing Small Stones flying about in Blasting, 35.—Quarry Shields, 36.

TAMPING	37

Different Materials employed, 37.—Experiments upon the Comparative value of Sand and Clay for Tamping, 40.—Experiments on Tamping with Packed Sand, 44.—Sand Tamping inferior, or worthless, 45.—Experiments with Tamping of Clay, broken Brick, and broken Stone, 47.—Tamping *Plugs*, 50.—Iron Plugs, 51.—Objections to Plugs, 51.—Pins and Cones, 52.—Iron Cones and Arrows, 53.—Experiments on Tamping with iron Plugs or Cones, 55.—Conclusions from preceding Experiments, 59.

TUNNELLING	60

Disadvantage of the work, 60.—Mode of Proceeding, 61.—A Good Record of Results of Working Tunnels much wanted, 62.—Assumed Advantages of constructing *Double* Tunnels, 63.—Denied, 64.—Sinking Shafts, 66.—Effects of Strata, 67.—Drainage, 68.—Ventilation, 69.—French Method of Ventilating Tunnels, 71.—Cost of Tunnelling, 72.—Cost of Road Tunnel at Drenodrohur, in Kerry, 72.—Cost of Railway Tunnel at Liverpool, 73.

APPENDIX	74

Mode adopted at Marseilles for enlarging inner End of Hole, 74.—Patent for ditto to Baron de Liebhaber, 75.—Blasting in the Jumna, and at Delhi, 76.—Blasting Plymouth Limestone, 77 and 79.—Quarrying the Granites of Aberdeen and Peterhead, 78.—Implements and Cost at Plymouth Quarries, 79.

SLATE QUARRYING	81

Geological Features of the Strata, 82.—Quarries at Bangor, 82.—French Slate Quarries, 83.—Quarries in Germany, Switzerland, Italy, &c., 83.—Slate hardened by Baking, 83, note.

FIRING MINES BY VOLTAIC ELECTRICITY	84

Removal of Round-Down Cliff at Dover, 84.—Charges, 85.—Chambers, Shafts, and Galleries, 85.—Voltaic Batteries, Daniell's, and Plate Batteries employed, 86.—Cartridges, 87.—Tamping, 87. The Firing and Results, 87.

CONTENTS. vii

PART II.

PAGE

STONES USED IN BUILDING 88

Division of Stones into *Granite, Slates, Limestones,* and *Sandstones,* 89.—*Granite,* 89.—Constituents, and Districts where found, 89.—*Herm* (Guernsey) Granite, 90.—Table of Results of Wear of different Granites, 92.—Table of Experiments on Crushing Granites, 94.—Experiments upon Irish Granites, 95.—Serpentine and Porphyry, 95.

Slate.—Constituent Substances, 96.—Districts where worked, 97. Irish Slates—Results of Experiments, 97.

Sandstones.—Chemical Analysis, 97.—Weight and Composition, 98.—Mode of using and selecting best description for Building Purposes, 98.—Examples of Durability of Sandstone Buildings, 99. —Experiments upon Crushing Sandstones, 100.

Limestones, 101.—Analysis of Oolites and Limestones, 101.— Weights of ditto, 101.—Composition of ditto, 102.—Analysis of Magnesian Limestones, 102.—Weights and Composition of ditto, 102.—Buildings of Magnesian Limestone, 103.—Buildings of Oolites and other Limestones, 103.—Summary of Results upon Sandstones and Limestones, 105.

DESTRUCTION OF BRIDGES IN SPAIN 107–114

Addenda by the Publisher—Attempts made at Pesth in Hungary, 115—117

Addenda—Attempted destruction of the bridges over the Ticino at San Martino, near Magenta and Buffalora, &c. . . . 117—119

Services of the Royal Engineers in the Crimea, by Capt. L. Vernon . . . . 120—127

QUARRYING AND BLASTING ROCKS.

In a "Rudimentary Series," such as the one to which this volume belongs, perhaps no subject can be introduced with greater fitness than that of quarrying or stone-getting,—certainly one of the rudimentary arts connected equally with the duties of the engineer, the architect, and the practical builder. Any attempt to recount the history of the art would oblige us to extend our retrospect to a very early date, perhaps we should rather say to a dateless period; to call to mind the structural wonders of Tyre, of Sidon, and of Thebes; and to close the history with the confession that, however the mechanical modes of wielding the wrought masses may have received improvement in modern times, the primitive methods and tools of the quarryman and of the mason have, if we except the introduction of gunpowder, become no more altered in our hands than have the chiselled works of the ancient sculptors been surpassed by the productions of later ages. The one powerful assistant, however, to which we have referred, has been enlisted in the service of the quarry within a comparatively recent period, and its aid has certainly supplanted to a great extent the action of the wedge and the hammer.

Without this valuable auxiliary, the work of separating large masses of stone from their native blocks or beds must have been infinitely more laborious, and have needed also a skill of no mean degree. The art of applying gunpowder to this purpose has, however, remained very nearly the same since the first experimenter "jumped" and "charged" a bore in the stubborn rock, which till then had been attacked only with the chisel and the wedge.

The implements employed for quarrying by mere manual labour are simply the sledge hammer or mallet, the borer or chisel, and the wedges, besides trucks, and such gearing as may be required to facilitate the removal of the blocks when detached. Of these, the only one which needs any description is the chisel, which is an iron rod with a steel cutting end, welded at one extremity, and flattened to a wedge-like form at the other. The implements used in blasting are of a similarly simple character, and consist of the borer or jumping tool, the scraper for clearing the hole of the chips produced, the claying bar for driving in dry clay if the hole be too damp for the immediate introduction of the charge, the needle, which is driven into the charge, and remains while the hole is filled up with stones, &c., so that, when ultimately withdrawn, a channel is preserved communicating directly with the powder. While the needle, which is a long thin copper rod, remains in its place, the space around it is filled up, by means of the tamping bar, with stones, &c.

In the following paper, which is from the pen of Major-General Sir J. F. Burgoyne, we have a most useful collection of facts upon the subject of blasting, together with a complete account of the process and the materials employed, and several hints for the improvement of both, which we trust will receive all the attention due to their high authority and intrinsic practical value.

The operations which constitute the entire process of blasting are, boring the holes, loading, and firing.

BORING THE HOLES.

The best means of expediting the operation of blasting, would be by any contrivance that would render the boring of the holes more quickly executed.

The ordinary implements used for this purpose are, the jumper or cutting-tool, the hammer, and the scraper.

There is much discrepancy in the account given in different places, of the time required for boring holes; arising from differences in the precise quality of the rock, the mode of working, and the different bases of calculation.

The following, obtained from John Mac Mahon, Esq., of Dublin, is the result of some considerable experience in quarries of granite of good quality at Dalkey, in the neighbourhood of Dublin.

" 3-inch jumpers, used in boring holes from 9 to 15 feet deep; 2 men striking, and 1 man holding and turning the jumper, bored on an average 4 feet in a day, or 5 feet with a 2½-inch jumper, which was frequently used for boring the same depth.

" 2¼-inch jumper, for holes from 5 to 10 feet deep; 3 men as above would bore on an average 6 feet each day.

" 2-inch —— holes from 4 to 7 feet deep; 3 men would bore 8 feet of such holes.

" 1¼-inch —— holes from 2 feet 6 inches to 6 feet; the 3 men bored 12 feet.

" In working the two last classes, a strong boy will answer to turn the jumper instead of a man.

" 1-inch jumper, for breaking the fragments of rock to smaller pieces; 1 man bored 8 feet per day.

" The waste of steel and iron was nearly as follows :—a 3-inch jumper took for its bit 2 lb. of steel, with which it would bore 16 feet, on being dressed or sharpened 18 times; waste of iron 18 inches to each steeling, or 1⅛ inch for each foot bored.

2-inch jumper took 1 ¼ lb. of steel.
1¼-inch   ,,   ,,   ¾ lb.
1-inch   ,,   ,,   3 oz.

" They would all bore from 18 to 24 feet with each steeling, and require to be sharpened about once for every foot bored.

" The weight of the hammers used in boring with each class of jumper was as follows :—

18 lb. hammers for 3-inch jumpers,
16 lb.   ,,   ,,   2½ and 2¼-inch,
14 lb.   ,,   ,,   2 and 1¼-inch,
5 to 7 lb. ,,   ,,   1-inch, used by one man.

" Churn jumpers, so called from the manner in which they are worked, from 7 to 8 feet long, with a steel bit at each end; general diameter, 1⅛ to 1½-inch. Two men would bore with them about 16 feet per day."

These last are much more efficient than those struck on the head with a hammer, and are sometimes used with a spring rod and line. They are applicable to holes that are vertical, or nearly so, and to rock that is not too hard: in the granite at Kingstown they were abandoned on account of the edges turning so fast, that the frequency of the necessary sharpenings gave the advantage to the use of the hammer:

where they can be used, the work will be performed at a far more rapid rate than even above mentioned. In boring artesian wells to great depths the application of the tool is entirely on this principle.

A still quicker mode of boring would be by *drilling* the holes, if it were feasible. It has often been proposed, but the cutting edges of the tools will not stand in any kind of stone.

Where charges have been exploded without blowing out the tamping,* it may be very desirable to bore the latter out for the purpose of introducing another charge; in such case the hole need not be of the size of that originally made, as 1-inch bore will be adequate at any time.

In clay tamping, a hole may be bored out with the jumper and hammer at the rate of about 26 minutes for 3 feet. Broken stone tamping will be bored out at the rate of 20 minutes for 3 feet.

A very few trials were made to bore a hole through clay tamping with an auger; the labour was less, as fewer men were necessary; the time consumed was somewhat more, but the tool was capable of improvement, and the men were new to its use.

It is conceived that augers, properly constructed, may be used to advantage in re-boring through clay tamping for successive blasts from the same hole, to the gain of a saving in time and labour, and avoiding the application of water, which is necessary with any kind of jumper.

In case of a miss-fire it is a very dangerous practice to re-bore the hole, and has occasioned very many bad accidents; it is very properly usually forbidden altogether. If the hole be vertical, or nearly so, and the needle or fuse hole can be cleared, so as to *ensure* a thorough wetting of the charge, by pouring water down, it might be done with safety; but sometimes the very object of the quarryman is to save the powder,—a very unworthy one for which to incur the great risk of killing or maiming two or three workmen.

To prevent the loss of any large charge in this way, a hole is sometimes bored on one side, and within a few inches of the one that has failed, to the same depth as its charge, and being loaded and exploded, has had the effect of igniting the other also.

* The "*tamping*" is the filling up of the hole in which the charge of powder, &c., is deposited.—ED.

An apparatus for boring to considerable depths has recently been introduced into this country from mines in Germany* by Charles Vignoles, Esq., C.E., which it is believed is far more efficient than any hitherto employed in Great Britain.

This machinery has been in operation, it is believed, on the Manchester and Sheffield Railway, of which Mr. Vignoles was the engineer; its direct application being to bore into and ascertain the precise qualities of the strata through which the great tunnel (3 miles in length) will be carried, which is to form the communication through the mountain ridge that divides the eastern and western inclines of England.

The principal feature in the process is that the cutting tools are attached to a heavy weight, and worked by a rope, instead of by rods with screw joints.

The rope is wound round a cylindrical roller by a winch, and there are several ingenious contrivances in the details and parts of the machinery, that tend to facilitate the operation.

Upon the judicious selection of the position of the holes will in a great measure depend the useful effect of the blast; but two leading errors are committed by quarrymen or miners in general, viz., selecting an injudicious position for the charge, by which the action of the powder is exerted in the direction of the opening where it was introduced; and the adopting as a rule for the several charges, to fill a certain number of feet or inches of the hole bored, usually one-third of its depth, instead of employing given weights adapted to the *lines of least resistance.*

The *line of least resistance* is that line by which the explosion of the powder will find the least opposition to its vent in the air. This need not necessarily be the shortest line to the surface; as, for instance, a long line in earth may, from the same charge, afford less resistance than a shorter line in rock.

Supposing the matter in which the explosion is to take place to be of uniform consistence in every direction, charges of powder to produce similar proportionate results ought to be as the cubes of the lines of least resistance, and not according to any fanciful depth of hole bored.

Thus, if 4 ounces of powder would have a given effect

---

* It is said that boring on a similar principle has been long used in China.

upon a solid piece of rock of 2 feet thick to the surface, it ought to require 13½ ounces to produce the same effect upon a piece of similar rock 3 feet thick; that is,

	Cube of 2-feet line of least resistance.		Charge of powder in ounces.		Cube of 3-feet line of least resistance.		Charge in ounces.
as	8	is to	4	so is	27	to	13½

or, what is the same thing, half the cube of the line of least resistance expressed in feet, will, on *this particular datum*, be the charge in ounces, as follows:—

Lines of least resistance in feet.	Charge of powder
	lb. oz.
1	0 0½*
2	0 4
3	0 13½
4	2 0
5	3 14½
6	6 12
7	10 11½
8	16 0

These quantities being of common merchants' blasting powder, will be found adequate for any rock of ordinary tenacity; but a precise datum should be ascertained by a few actual experiments on the particular rock to be worked.†

Thus, with a 2-feet line of least resistance (A to B, fig. 1), whether 4 ounces or 6 ounces, or 8 ounces, are requisite to produce a good effect; with 3-feet line of least resistance, whether 13½ ounces, or 18 ounces, or 27 ounces, &c.

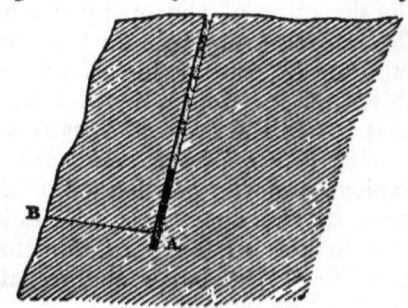

Fig. 1.—(Section.)

---

* To so small a quantity as ½ ounce a little excess might be added, but ¼ ounce, or ⅛ ounce more, will be sufficient.

† In the granite quarries of Kingstown (near Dublin) these charges were sufficient to open the rock where there were no fissures, apparent weakness, or other advantage; but where the line of least resistance was not that of the hole bored, the effect was either to bring it down, or only to crack it, according to the quality of the powder used.

On the results of these trials a scale may be adopted for guide in the service.

With regard to the first error above mentioned, that of leaving the action of the powder to be exerted in the direction of the hole bored, one consequence is that, with small charges, commonly a part of the explosion finds comparatively easy vent by that opening (in spite of the best tamping), and is wasted; and it is only the excess that acts in producing the effect required on the rock; whereas, if the explosion were forced through another direction, the whole of its power would be exerted beneficially. But the great objection is, that the rock is then so much more firmly bound all round the charge, as to oppose and lessen in a very great degree the extension of the effect of the explosion.

It must be understood that, even although the line of least resistance should be in the direction of the hole bored, the *depth of the hole* will by no means be the measure by which the proportions of powder for the charge can be taken according to the above rule, namely, as the cubes of the lines of least resistance.

1st, because the tamping, however good, or by whatever contrivance strengthened, cannot be equal in strength to the solid rock.

2ndly and chiefly, because of the various proportions of the entire depth of the hole occupied by different charges of powder: thus, $\frac{1}{2}$ ounce of powder will occupy an insignificant proportion of the depth of even 1 foot of a 1-inch hole, and also the 4 ounces for a 2-feet line of least resistance would fill only 2 inches of a 1-inch hole, and consequently occupy one-twelfth of the 2 feet, and leave 1 foot 10 inches of tamping; while the $13\frac{1}{2}$ ounces for a 3-feet line of least resistance would occupy of a 1-inch hole above 6 inches, that is, more than one-sixth, and leave only 2 feet 6 inches of tamping, and consequently of resistance, such as it is.

This might be remedied in one way, by applying holes of larger diameters for increasing charges, but, by so doing, an increased amount of the less resisting medium (the tamping) would be the consequence, which again renders the calculation, founded on a resistance of solid rock, incorrect.

There is another reason why the powder is ill applied when the explosion takes place in the same line as the bore, which is, that it is placed longitudinally with the line of least resistance, as at c, in fig. 2, and not perpendicular to it, as at P; when much extended, the elongated form in either is bad, but it is worst in the former case.

Fig. 2.—(Section.)

When the common mode of blasting is adopted, a loud report is heard like a gun (louder in general in proportion to the less *useful* effect produced), and fragments of stone are frequently thrown to a considerable distance; but when done judiciously, the report will be trifling, and the mass will be seen to be lifted, and thoroughly fractured, rent, or thrown over, without being forcibly projected.

It is the irregularity and extent of the violent explosion following the ordinary process, that renders it so peculiarly difficult to form an accurate judgment on the proper charges for each circumstance; the consequence is a practice purely empirical. The miner or quarryman will give as his rule either a proportionate depth of hole, or, aware how frequently that must prove erroneous, is driven to his usual answer, that he knows from the *appearance* of the situation, what to apply; that is, in fact, admitting that it is a matter of caprice, and provided a certain effect is produced, he is little aware how much time, powder, and labour, may have been wasted.

It is difficult to make ordinary quarrymen, or even overseers, understand correctly the meaning of the lines of least resistance: after appearing to comprehend it, they are frequently observed to confound it with the length of hole bored, or with some conceived necessary direction either vertical or horizontal.

With respect to the second error mentioned, it can easily be shown how very erroneous *must* be the rule of measuring the charge by any given proportion of the depth of the hole, since the quantity of powder will in that case depend, not only on the depth, but also on its *diameter*, with which it can have no relation: thus, if a 6-feet hole is to be bored, it may be an act of chance or caprice whether a jumper of 1½-inch gauge, or of 2 inches, or of 2½, were used; whereas the third of the depth, or any given number of inches of the 2-inch, would hold very nearly *double* the quantity of powder that would be contained in the 1½-inch; and of the 2½-inch, one half more than in the 2, and nearly *three times* as much as in the 1½.

Such a rule also takes no account of the quality of the rock, which in reality will cause the greatest difference in the effect; a given depth of hole being applied to hard or soft rock indiscriminately.

Although some allowances may be made in extreme cases, yet it will be found in most books and papers on blasting rock, that a *usual* charge is one-third of the depth of the hole; and the same will be found to be the actual prevalent practice.

As to the experience by which it might be assumed, that the miners will modify this rule, and regulate the proper charges to give in each case, the value of such practice, unaided by better *principles*, must be small, where the results are so indistinct. A loud explosion takes place, and the rock is more or less separated, but no proof whatever is afforded that the charge has been precisely, or even nearly, what it should have been; and being regulated by no rule (for in this case of leaving it to the miner's judgment, the only rule is abandoned), the experience, to be valuable, should be of precisely similar cases, whereas in blasting they are constantly varying, in size and depth of hole, and in many other circumstances. If the rock were uniform, and the application of the charge always in similar holes and situations, a tolerable rule of thumb experience might perhaps be obtained; but it is quite otherwise; and among the circumstances that must tend to perplex an ordinary miner workman, would be, that the true principle for charges is, as the *cubes* of the thickness of the resisting medium, but which he would certainly regulate by a much more gradual proportionate increase, such as doubling, trebling, or squaring at most.

In order to quarry with good effect for saving labour and powder, an exposed front, either vertical or horizontal, should, if possible, be established on the rock on which to operate, principally to obtain a line of least resistance in a different direction from that of the hole bored.

Thus, if a charge of powder were placed at P, fig. 2, with a line of least resistance to B, the explosion would force its way through at B, shatter and loosen the whole mass at D, and make cracks to a great extent towards E; whereas if the hole had been bored direct from B to P, or as at C (as is usually practised by common quarrymen), the resistance being excessive in every direction, except in the direct line of the hole bored, it may be easily conceived that the same charge would produce far less effect.

Or, to adopt a more practical illustration, suppose ledges of rock require to be cleared away to a certain level for a road, navigation, or other object; instead of boring holes + + +, the effect would be far better by inclined holes, A, B, C, fig. 3, applied in succession after the above-mentioned principle.

Where there is a high face of rock, a system of undermining may be advantageously employed: thus a blast at A, fig. 4, would make an opening easy from C to D, and the mass E, if not shaken, which it probably would be, so as to be worked on with crowbars and wedges, would be brought down by very slight subsequent blasts.

When the rock is stratified, and in close parallel beds and seams, the holes should be bored in the direction of the joints, and the powder laid along them as at A, fig. 5, which will have much more effect in lifting large masses than if placed across the grain, and the operation of boring will be easier.

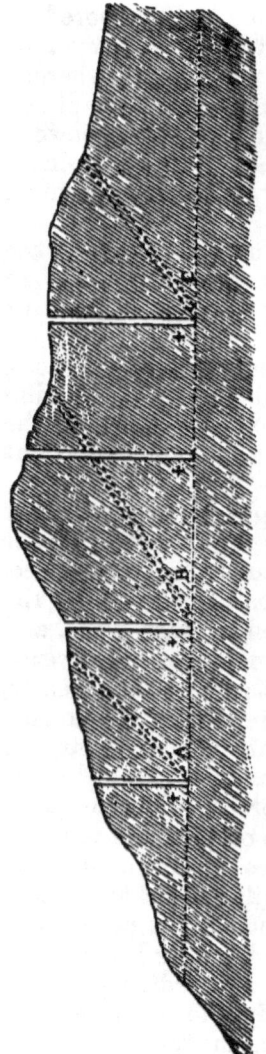

Fig. 3.—(Section.)

The worst situation for a charge of powder is in a re-entering angle, as at A, fig. 6: the rock exerts such

QUARRYING AND BLASTING ROCKS.

pressure all around it, that very little effect can be expected nor is the position much improved at B.

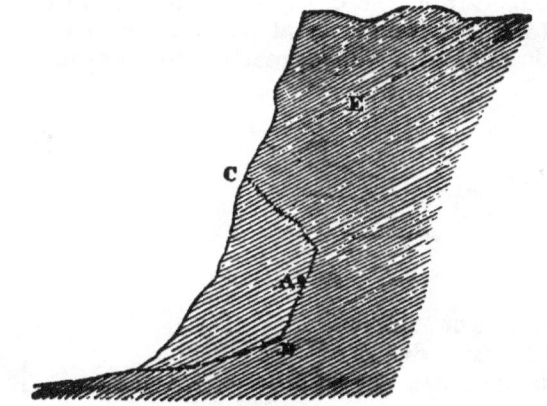

Fig. 4.—(Section.)

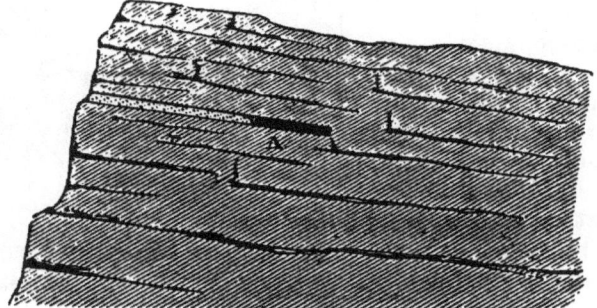

Fig. 5.—(Section.)

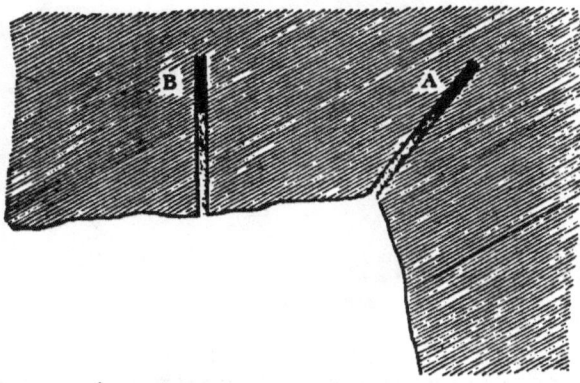

Fig. 6.—(Horizontal Section.)

This situation of a re-entering angle occurs very frequently and should be avoided as much as possible. Thus a charge may be lodged in a hole C, fig. 7, having the same length of line of least resistance as at other holes, A or B; but the effect of the explosion will be greatly reduced by the masses D, E.

Fig. 7.—(Plan.)

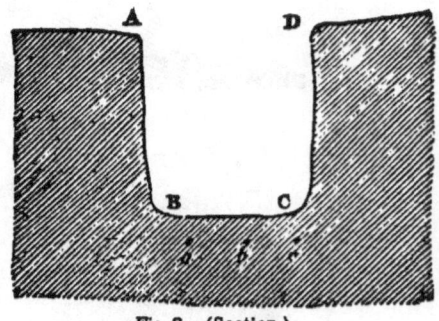

Fig. 8.—(Section.)

An important illustration of this disadvantageous position for the charge will be experienced in cutting through any narrow confined space A, B, C, D, fig. 8, either horizontally, as in the first drift, or opening, all through a tunnel, or vertically, as in sinking a shaft: blasts at *a*, *b*, or *c*, must be extremely ineffective.

A projection, on the contrary, is the most favourable situation to produce the greatest effect with the smallest

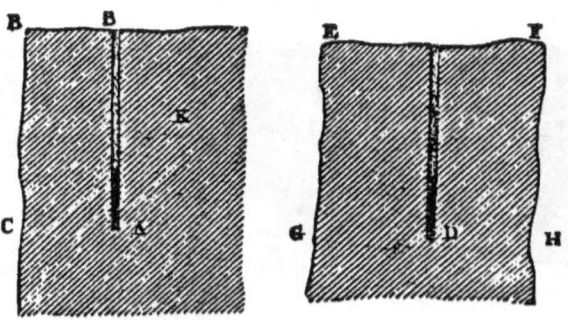

Fig. 9.—(Sections.)

means: a given quantity of powder, for instance, at A, fig. 9, would remove the mass B, B, A, C, and make partial cracks on the side K, but the same quantity at D would remove E, F, G, H, or nearly double the mass*.

Cases, however, frequently occur requiring a deviation from what, under ordinary circumstances, would be the most favourable application of the charge, either owing to the quality of the rock, or where other objects are of more importance than the consumption of powder, or of labour in boring.

The rock may require to be cut to a particular form, as, for instance, when preparing it to receive foundations of masonry; or certain blocks of the stone may be required of particular forms or dimensions: an excess of powder may be applied to increase the shock for bringing down any loose mass in a peculiar state, or *vice versâ*, smaller proportions may from circumstances be sufficient to produce as much effect as is required: these irregularities are very frequent, but it is not the less necessary to understand the correct principles, and not to be carried away with the idea that the whole is a mystery.

It may also be urged that there are cases where the system of working on a line of least resistance, different from that of the hole bored, cannot be followed; such, for instance, as in sinking a shaft, or cutting the first drift-way of a gallery or tunnel.

This is very true, and must cause such operations to be peculiarly costly and slow; and though the rule above recommended for regulating the charges would be inapplicable, still that of taking any proportionate depth of the hole would be quite erroneous.

The system to be adopted in such instances should be to apply the mode of tamping that would give the greatest possible resistance, and to endeavour to obtain by trials the amount of charge by weight that will barely disturb such tamping; in this way the full effect of the powder will act upon the rock, and where that is not very great, a second shot from the same hole will be sure to be very decisive.

Among the cases not admitting of fixed rule, and where a great deal *must* depend upon the intelligence and experience of the directing quarryman, is that of irregularity of joints, or seams.

* It will not have quite this effect, as the greater resistance on the side K, in the one case, will increase the effect towards C; but that circumstance does not affect the general consideration of the principle here adverted to.

The following instance will explain this:—

A hole, A, B, fig. 10, of 4 inches diameter and 16 feet deep, was bored to the back of a projecting mass of granite, and 6 feet above a natural slide joint, as shown in the plan and section.

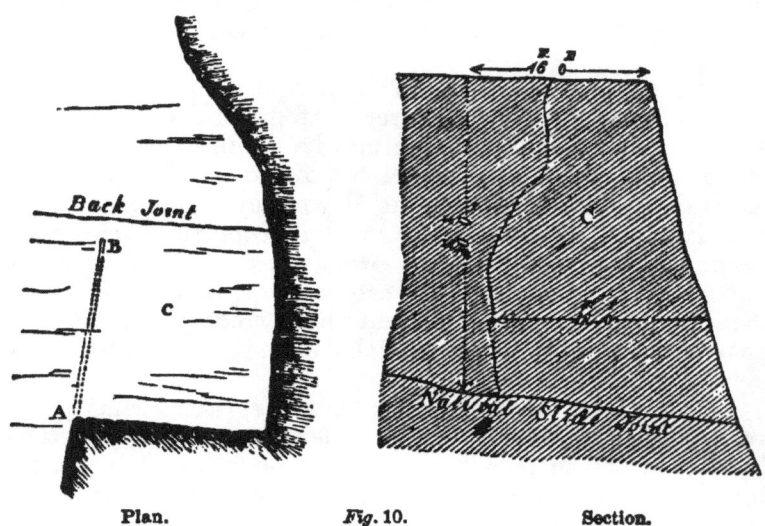

Plan.   Fig. 10.   Section.

Had there been no projection, nor any joint to afford an advantage, it would have required probably 182 lb. of powder to break through a resistance of 18 feet; but as it was circumstanced, between 35 and 36 lb. (occupying 7 feet of the hole) broke off and overturned an enormous mass C, cutting it down as shown by the dotted lines: the fragments of course were large, one piece containing 80 cubic yards; and were very appropriate for cutting into ashlar of large dimensions.

Another instance, tried at Dunmore East, near Waterford, may be given of the force of the action of powder, even in an open joint: the same experiment also incidentally serves to illustrate another object of inquiry.

The rock was a very hard conglomerate, in fair parallel beds; the surface was even; the side H, H, fig. 11, had been already excavated; there was a joint or bed, parallel to the surface, and 7 feet below it. A hole a, of 2 inches diameter and 6 feet deep, had been loaded with 1 lb. 14 oz. of powder, (15 inches), and made a straight crack $c\ c$, and to k, of 14 feet long, and down to the bed.

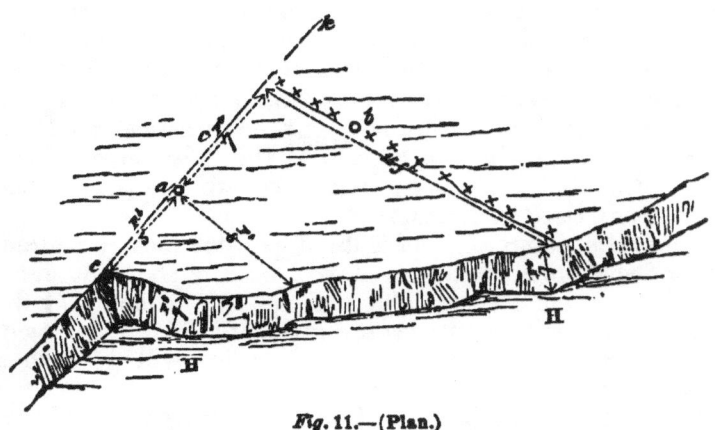

Fig. 11.—(Plan.)

A trial was then made, whether openings might not be directed to particular lines: 13 plug holes were drilled along the line + + + 8 inches deep and 1¼ inch diameter; a 2¼-inch hole was then bored at *b*, to a depth of 5 feet 10 inches, and loaded with 4 lb. (2 feet) of powder: when fired, the line was opened very nearly along the line of the plug-holes; the separation did not exceed half an inch.

The hole *b* was then cleared out, and loaded with 8 lb. of powder, and the explosion sent the whole mass forward upwards of 2 feet without any new fracture; the cubic contents of the mass being 663 feet, weighing about 51 tons.

In most extensive quarries of stone, much of the practice must depend upon the intelligence with which advantage is taken of the position and nature of the joints and fissures: still many errors are committed from a want of knowledge of the best application of powder to a perfectly solid mass; and in cases where the mass to be removed is small, or the openings to be made, confined, but little advantage can be gained by the joints, and the application should be chiefly to given charges for lines of least resistance.

The advantage obtained by joints is one reason for rather reducing the calculated amount of charges, particularly in large explosions; because, although nothing that is not perceptible can well *augment* the force of resistance, fissures or joints that may not be seen on the surface may have the effect of *reducing* it.

It would be of much advantage, in many cases, if the powder could be placed in a more compact form, than occupying a considerable length of a hole of comparatively small diameter: the position it must assume in these holes is generally unfavourable for producing the best effect, and in some cases renders it impossible to apply so large a charge as would be desirable; but no practical mode of enlarging the space at the bottom of the hole has yet been contrived, except perhaps by successive explosions from the same, as practised at Gibraltar.*

It may be assumed that 1 lb. of powder loosely poured but not shaken or compressed, will occupy about 30 cubic inches; a cubic foot weighs consequently aboot 57½ lb., although different quantities are given in different tables of specific gravity: if close shaken, powder will go into a smaller compass.

The following Table from Colonel Pasley's (now Major-Gen. Sir Chas. Pasley, K.C.B.) Memoranda on Mining will give the means of calculating the space occupied by any given quantity of powder in round holes of different sizes from 1 to 6 inches:—

Diameter of the hole.	Powder contained in one inch of hole.		Powder contained in one foot of hole.		Depth of hole to contain 1 lb. of powder.
Inches.	lb.	oz.	lb.	oz.	Inches.
1	0	0·419	0	5·028	38·197
1½	0	0·942	0	11·304	16·976
2	0	1·676	1	4·112	9·549
2½	0	2·618	1	15·416	6·112
3	0	3·77	2	13·24	4·244
3½	0	5·131	3	13·572	3·118
4	0	6·702	5	0·424	2·387
4½	0	8·482	6	5·784	1·886
5	0	10·472	7	13·664	1·528
5½	0	12·671	9	8·052	1·263
6	0	15·08	11	4·96	1·061

In practice, the holes are somewhat irregular; this table, however, will be sufficient to ascertain, nearly, the depth required for any charge.

* In the Appendix will be found a memorandum by General Burgoyne on the use, at Marseilles, of nitric acid for the purpose of enlarging the inner end of the hole; and also a brief description of a patent obtained by the Baron Liebhaber for an apparatus for this purpose.—Ed.

OF THE POWDER AND THE CHARGE.

Gunpowder used for the blasting of rock is notoriously of inferior strength to that sold for sporting, or manufactured by Government for the army and navy; and there is an impression (I believe nearly universal) that it is right that it should be so, not merely because pound for pound it is cheaper, but because it is thought to be positively better for the object, on account of its less rapid ignition, and assumed quality of giving what the miners call a *heave*.

This opinion appears to be founded on a fallacy.

Inferior powder *cannot* be used in war, or for sporting, without the disadvantage being immediately apparent; while in blasting it can be made to answer the purpose: this, with its comparative cheapness, has led, no doubt, to its being introduced and constantly made use of, without much investigation as to the policy of its employment in preference to a material of superior strength.

The argument used in its favour is, that, by igniting slowly in comparison with the other, the power is more forcibly and efficiently applied for the required object, than by the rapid shock of the superior powder, such as is undeniably requisite for impelling projectiles.

This reasoning would seem to imply that the rock will be opened better by a force of *pressure* than by that of a sudden *shock* or *blow;* which, however, may be disputed, even supposing, what is probably not the case, that the elastic vapour generated by either is the same. Rock being of a brittle nature, it is reasonable to suppose that the sudden violent shock would make more extensive cracks, which is the great object, than a more slow action.

The following are the observations I have been able to collect on this head; and they tend to confirm the impression of the good policy of employing stronger powder for blasting, even at increased prices: more research, however, would be necessary to establish the fact entirely, and to fix the relative value of each gradation in quality.

Having procured from great contractors and respectable dealers eleven samples of Merchants' blasting powder, stated to be that of the principal manufacturers, they were analysed, proved with the éprouvette mortar and éprouvette gun, and compared by the bursting of shells: the results will be seen in the annexed tables.

Qualities and Proofs of eleven samples of Merchants' blasting powder, as compared with Government cannon powder.

Number of samples.		Results of analysis per centage.				Ranges from éprouvette mortar.			No. of degrees and tenths by éprouvette gun.	Remarks.
		Nitre.	Sulphur and Charcoal.	Loss.	Total.	1st.	2nd.	Mean.		
		grs.	grs.	grs.	grs.	ft.	ft.	ft.	degrees.	
1	Said to be of same manufacturers, but procured from different dealers.	72½	25	2¼	100	97	93	95	16·6	Highly impregnated with foul salts.
2		66	32½	1½	100	142	137	139½	17·3	Contains foul salts, but not so much as No. 1.—Deficient in nitre.
3		66	32	2	100	150	91	120½	18·7	Deficient in nitre, and highly impregnated with foul salts.
4	Two qualities, from the same manufacturers, but procured from different dealers.	66½	31½	2	100	79	99	89	14·6	Same as last.
5		75	24	1	100	125	148	136¼	...	Nitre very impure.
6		73½	25½	1	100	83	67	75	...	
7	Two qualities, same manufacturers.	73½	24½	2	100	118	113	115½	17·7	Highly impregnated with foul salts.
8		66	32	2	100	43	55	49	16·9	Do., and deficient in nitre.
9	Three qualities, same manufacturers.	73	25½	1½	100	169	158	163½	12·0 to 15·2	Contain foul salts, but by no means so much as the preceding samples.
10		73	25½	1½	100	128	148	138		
11		73	25	2	100	127	107	117		
	Good Government cannon powder.	75	25	...	100	...	...	265	21·0	Ingredients pure and very intimately mixed.

The best proportion of nitre (the most valuable ingredient) is 75 per cent.

The éprouvette mortar is 8 inches in diameter, and is charged with 2 ounces of the powder, and an iron ball of 68½ lbs. weight; the Government good cannon powder gives an average range of 265 feet. The Government powder somewhat deteriorated, and reserved for blasting, gives a range of 240 feet.

The éprouvette gun is of brass; its bore 1¾ inch in diameter and 27·6 inches long; it weighs 86½ lbs., and is suspended from a frame: being charged with 2 ounces of the powder, without any shot or wadding, it is fired, and the extent of the recoil is measured by an index on a graduated arc.

The éprouvette gun is considered to be rather adapted to try the strength of *fine grained* powder than of the coarse; the fine grained Government Rifle powder will give 25 or 26 degrees; the Government good cannon powder 21 degrees; that tried at the same time with the above, 20·5. It is extremely probable that in many instances of these coarse grained qualities, and of the Merchants' powder, igniting slowly, much of the charges may have been thrown out from the gun in each case unignited, and perhaps in some degree from the mortar.

Some discrepancies will be observed in nearly all proofs of gunpowder, but rarely to the extent that will be noticed in this Table: they show, however, how very unequal may be the qualities of the article as obtained at different times, from different dealers, and subject to a variety in their condition from modes and time of keeping; and they also exhibit a very strong presumption of universal inferiority.

Experiments on the relative strength of Government cannon powder and Merchants' blasting powder, by the bursting of 5½-inch spherical case shells.

	Government Cannon Powder.				Merchants' Blasting Powder.			
No. of experiments.	No. of shell.	Charges of powder.	Effect.	Observations.	No. of experiments.	No. of shell.	Charges of powder.	Effect.	Observations.
		oz.					oz.		
1	1	4	None.		4	2	8	None.	Loaded and fired warm from previous explosion.
2	2	6	None.		5	3	10	None.	
3	1	8	Burst.	Loaded and fired warm from previous explosion.	6	2	12	Burst.	Third trial of same shell, loaded and fired warm.
8	5	7	None.		7	4	11	None.	
9	6	8	None.		12	9	12	None.	
10	7	9	None.		13	10	14	Burst.	
11	8	10	Burst.		14	11	13	Burst.	
15	12	9	Burst.	Second trial of same shell, but quite cold.	17	4	12	None.	Second trial of same shell, but quite cold.
16	3	8	None.	Second trial of same shell, but quite cold.	19	6	12	Burst.	Second trial of same shell, but quite cold.
18	5	9	Burst.						

The comparative strength by this mode of trial would seem to be about 9 parts of the Government, equal to 13 of the Merchants' powder.

It is very difficult to obtain any very precise results from trials on the rock itself: the effects vary so greatly under circumstances to all appearance precisely similar, that we are driven to reason very much from analogy; and where the blasting is judiciously performed, the tamping but slightly, if at all removed, and the rock merely opened and not violently ejected, it certainly would appear reasonable that the powder showing such superiority of strength in the above-described experiments would be by far the most efficient in its action on the rock.

A few trials that were made at Kingstown to elucidate the point seemed to prove the truth of this reasoning.

Charges calculated on the basis of $\frac{1}{2}$ ounce for 1 foot, and augmented in proportion to the cubes of the lines of least resistance, were exploded in a high face of very solid rock; and the result in every case was very marked in favour of the Government powder, even to the conviction of the miners present, who had previously expressed doubts on the subject.

Although the Government powder was applied to the cases which seemed to present the fewest advantages, the effect was decidedly superior, to the extent of usually dislodging a mass of rock; whereas the Merchants' powder in no instance did more than make cracks and fissures.

If the truth of this suggestion be acknowledged, the following advantages would attend the use of the superior powder.

1. As smaller quantities would go farther, the stock for consumption would be easier to stow away and to carry.

2. Greater effect would be produced with a smaller amount of labour, and, what is of more consequence in many cases, of time in boring holes.

3. By occupying a smaller space in the bottom of a hole, an increased resistance in the tamping would be obtained by its greater proportionate extent.

4. The Government powder, and the superior kind made for sporting, (the former in particular,) are *much less* subject to deterioration from keeping, than the ordinary blasting powder; this would effect a very desirable improvement, but it is not an absolutely necessary consequence of their being stronger, because the best preserving powders are not always the strongest.

According to Dr. Ure's Chemical Analysis, there is not

much difference between the mixtures of the Government Waltham Abbey powder, and those of the *first class of sporting* powder of the private manufacturers; the Government powder, however, resisted rather the best the hygrometric influence, that is, would absorb less atmospheric moisture, and consequently be best for keeping.

In the works carried on by the Royal Engineer department, the powder is usually from the Ordnance stores, sometimes being perfectly good, or even if deteriorated to the degree for its being appropriated to blasting, it is still much stronger than the Merchants' blasting powder.

Fine grained powder made with very superior care, and at superior cost, is manufactured for the Rifle Service by Government, and for shooting by private manufacturers: but it would be too costly in proportion to its increased superiority, and some of its properties not being necessary for blasting, it is considered that the best *cannon* powder would be the most advantageous to employ.

Blasting powder is sold by dealers in the country at from 50s. to 75s. per 100 lbs.; while nearly as good powder of this nature as can be made, such as the Government cannon powder, might be sold *by the manufacturers* at between 50s. and 60s.: supposing, therefore, that the cost, including the removal, the *dealers'* expenses and profit, should be one quarter or one third more than at present, the question will be, considering the advantage of using this superior kind, and the proportion which the cost of the powder bears in the general expenditure of blasting, how far, and under what circumstances, it might be desirable to incur that increased expense, making allowance at the same time for the smaller quantity that would be consumed.

Founded on the same reasoning of the advantage of more gradual ignition, and almost leading to the assumption that the blasting powder in its present state is even *too good*, is the assertion that will be found in all books on the subject, namely, that a mixture of fine and dry sawdust of elm or beech with the powder, in the proportion of $\frac{1}{2}$ of sawdust for small charges, and $\frac{1}{3}$ for large, will produce as good results as equal quantities of powder alone.

It is not assumed that the effect is produced by any decomposition of the sawdust, but simply by giving a little more space, and by dividing the particles of the charge, causing them to ignite more gradually, and thus to act with greater force on the rock, than by the more sudden explosion.

No account is given of any defined experiments tending to prove this fact; on the contrary, every trial affording positive results is against it: such a mixture has been tried in guns, and produced no useful effect whatever; and though of very simple application, it does not appear that in any place there has been a continued use of it.

There is indeed a deception in the first instance in the supposed proportions; for a mixture of two equal measures of the two ingredients, the sawdust being, as required, very fine and dry, and the powder of the usual large grain, will not fill above 1¾ of the same measure; consequently, two measures of the mixture will contain nearly ⅛ more powder than calculated upon: thus if two measures, each capable of containing 8 ounces of powder, be filled with the mixture of equal measured proportions, the quantity of powder will be nearer 9 ounces than the 8 calculated upon.

Altogether I feel little doubt of the application of sawdust being of no real value.

Another theoretical refinement, that is to be found in all works on blasting, is, that if a hollow space be left adjoining the charge, a much greater effect will be produced, provided always (and it is essential to bear it in mind) that the tamping be as substantial, and to *as great* extent, in the one case as in the other.

Thus, in two holes of similar dimensions, the charge c, fig. 12, with a hollow space D over it, will produce as good an

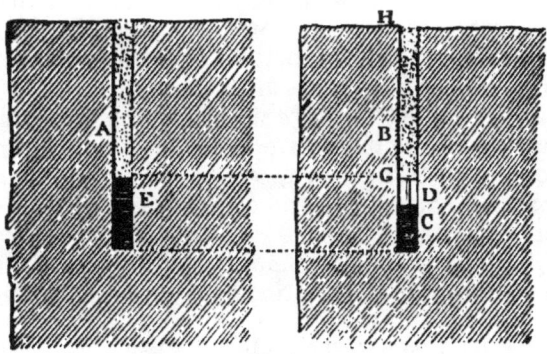

Fig. 12.

effect with ½ or ⅔ the quantity, as at the full charge E fully tamped, provided the tamping B, from G to H, be as good and as deep as that at A.

An increased effect will certainly be produced by such

hollow, in the same manner as with guns which are frequently burst by the occurrence of a hollow between the powder and the shot, but there must be great reason to doubt its *practical* utility: no accounts are given of the well-defined result of actual experiment, nor are any rules attempted to be laid down for the extent of the hollow spaces in proportion to the quantity of powder in the charge, &c., to produce useful effect; and yet these must be matters of consequence; nor is it anywhere stated that it has been ever practically continued to be used, notwithstanding the great saving of powder professed to arise from the adoption of this principle.

In large charges, the space that could be left would be too small to produce any useful effect; and in small charges, the more simple, quick, and cheap way, would be by using the full charge of powder.

This and the sawdust are among the refinements adverted to in books (on this as on many other subjects), but seldom, if ever, put in practice.

Another mode of improving the power of the charges of powder has been employed, it is said, in America. It is the mixing of a quantity of quicklime in a proportion of about $\frac{1}{4}$ with the powder, on the principle that it will absorb any little moisture in the powder, and itself produce some additional vapour in the explosion. It is stated, however, that it must be used soon after the ingredients are put together, it having been remarked that if left mixed for a whole night the powder was deteriorated, owing, as imagined, to the impurities of the saltpetre of which the powder was composed.*

With reference to the quantities of powder to be employed in blasting, different systems are adopted.

In quarries worked for large stones, and in great quantities, sometimes very large blasts are considered advantageous.

In those near Kingstown, where the granite stones for ashlar work are squared by the contractors to from 40 to 60 cubic feet each, 50, 60, and 70 lb. of powder were frequently exploded in a single blast, sometimes filling two-thirds of a hole of 4 or $3\frac{1}{2}$ inches diameter, and perhaps 20 feet deep.

---

* There are other reasons against this practice. The absorption of water must be attended with the evolution of heat; and it is highly probable that, under such circumstances, there would be a reaction between the sulphur and nitre, in the presence of so powerful a base, and the ultimate production of sulphate of lime. Whilst then, the simplicity of the operation is diminished, the powder is deteriorated; and, if much moisture be present, not without danger.

They have been applied generally where a projection of considerable height or length, showing joints in large features, offered a prospect of bringing down some enormous mass; and in this way they were usually very successful.

To give an instance of one:—

The hole was 19 feet 7 inches in depth, and 5½ inches diameter: the charge, 75 lb. of powder, filled 8 feet 10 inches of the hole, having consequently 10 feet 9 inches of tamping.

The mass that was brought down, or thoroughly shaken and rent, measured, on a rough calculation, 1200 cubic yards, or 2400 tons. The cost was calculated at £6 15s. 8d., thus:

	£	s.	d.
2 men, 14 days, at 1s. 8d. each	2	6	8
1 ditto, 14 days, at 1s. 6d.	1	1	0
75 lb. powder	2	0	0
Fuse	0	2	0
Smith's work, iron, steel, &c.	1	6	0
	£6	15	8

Of course there was a great deal of after-work and small blasts required for the separation of the large masses of which this shaken rock consisted, and reducing it to manageable shapes and sizes; but the work was greatly facilitated by this first effect.

At Gibraltar, the military miners under the Royal Engineers work on quite a different system.

The rock there is a peculiarly hard lime-stone marble: to bring down large masses, they bore their holes usually to about the depth of 9 feet with 2½-inch jumpers, and load them with about 4 lb. of powder; the explosion has no apparent effect, but the rock is shaken below: the needle hole is cleared out, and the hole again filled, when it will take perhaps from 8 to 12 lbs., and is fired again; a third charge, with perhaps 20 to 30 lb., is fired in the same manner, and sometimes a fourth, till the rock is very greatly separated and rent to the extent of 10, 20, or 30 feet in different directions. If the needle hole or tamping be deranged before the final explosion, it is bored out and re-tamped.

Under all ordinary circumstances I should much prefer the principle of this system to the other; it is more gradual and systematic, would require less labour in boring, and is less subject to waste of powder, or the violent projecting of stones.

In some rocks it may be liable to one objection, which is, the chance of any of the preliminary explosions tapping a spring or vent for water into the hole; no such springs are found at Gibraltar, but they are at Kingstown.

If the holes should not be vertical, or nearly so (which it seems they always are at Gibraltar), the tamping must be bored out at each explosion, to enable the next charge of powder to be introduced.

The prevailing fault in blasting is the using too much powder.

If the tamping be not blown out by the smaller charges, a very useful effect will have been produced on the rock by every explosion, even although the rock be not *apparently* much affected; the tamping will be easily bored out sufficiently to admit a fresh charge, which, being introduced and re-tamped, will be found to be very efficient.

The object is generally to loosen and bring the rock down in large masses, and not to shatter it into fragments: even for small stone, such as for road metalling, &c., it is better to bring out first large masses, and subsequently to divide them, either by small blasts with powder, or by crowbars and wedges.

OF LOADING.

The ordinary manner in quarries of loading and firing the holes that have been bored is,

I. To dry out the bottom, if necessary, with little wisps of hay.

II. To pour in the powder till it fills a certain number of feet or inches of the bottom of the hole. If the hole be vertical, or very nearly so, the powder will drop in pretty clear to the bottom; but if it be on an inclination, and not very steep, the powder must be scraped down, professedly with a wooden ramrod, but frequently with an iron scraper. If the hole be horizontal, a scoop is used, which is open at top, and by being turned round at the end of the hole leaves the powder there. If the hole incline upwards, a cartridge is employed.

III. A needle is then introduced, the point of which is l well into the charge of powder, and the top with a handle or eye extending to the outside of the hole.

IV. A little wadding, either hay, or straw, or turf, is inserted over the powder.

V. The tamping over the wad is very generally of the small fragments of the quarry stone and its dust (unless there be in it flint or other substance that *notoriously* strikes fire, in which case broken brick is commonly used), rammed down, by one or two inches at a time, by means of an iron rod or tamping-bar; the needle is frequently turned to prevent its becoming fixed.

VI. The last inch or two is filled with damp clay.

VII. The needle is carefully pulled out, and the opening it has left is filled with loose fine-grained powder; or with a long series of connected straws filled with powder, into the upper end of which is inserted a small piece of touch-paper that will burn about half a minute, which is lighted and communicates the fire.

The touch-paper is made by the quarrymen themselves, by soaking coarse paper in a strong solution of saltpetre or gunpowder, and then drying it.

The following is an expressive account of the process, as written by a quarryman of much experience:—

"The method of blasting is in every place nearly the same, as far as I have been able to make observations, and I have had charge of such work in Scotland, England, Wales, America, and, lastly, in Ireland. Different quarrymen may, it is true, not agree in everything; for instance, some prefer a small piece of dry turf for a wad over the powder before they commence ramming; others prefer hay or straw; but in ramming every one uses the same kind of stuff, that is, small pieces of any stone (that has no flint in it) that will go into the drill hole; but, in deep holes, ramming sand will do as well as anything.

"The usual method of blasting is very simple, and is as follows: first, to drill the hole, say 3 feet deep, $1\frac{1}{4}$-inch bore; in common cases, 6 inches powder will be sufficient in the bottom; then in with your needle, then your wad of turf or hay, &c.; then two or three blows with your rammer, and then in with a handful of small stones; four or five blows more of your rammer; and so on till you fill up the vacuity above the powder within 1 inch of the top; then fill said 1 inch with a bit of moist clay (not too wet), and then extract your needle; lastly, fill up the needle hole with fine powder, or, what is safer, put down straws filled with powder, and apply your match, to which set fire, and run as fast as you can, and you have the whole of it."

Nothing can be more, 1st, uncertain,—2nd, wasteful,—3rd, dangerous.—4th, unscientific, than the whole of this process.

1st. Missing fire occurs frequently,—some obstruction will arise in the needle hole; any moisture in the hole, or wetness in the atmosphere, will affect so small a quantity of powder as composes the train; loose powder for the train cannot be introduced into any but such as are vertical or very steep, and straws of any length are not easily passed through to the charge. All these and other circumstances must create much uncertainty.

2nd. To say nothing of the guess-work in the proportion of the charge, there must be much waste in the manner of introducing the powder, and in the occasional missing fire.

3rd. With regard to the danger, it manifestly pervades every step from the first handling of the powder.

4th. No rule is adopted for the charge, nor for the size of the holes, nor for the depth of tamping; no knowledge acquired of the best material and mode of tamping, nor any contrivances for accelerating and simplifying the process, or reducing its danger.

To obviate in some degree these defects, the following proceedings have been adopted with success:

To allot every charge by weight, according to a scale adapted to lines of least resistance, or to the circumstances of the case.

The overseer (or, if the work be sufficiently considerable, an express powderman) to have on the spot a strong copper canister containing from 3 or 4, to 10 or 12 lb. of powder, with a large mouth or opening, but thoroughly secured by a well-fitted cover from spilling, accident, or weather, and with a lock and key.

He should have a set of marked copper measures that will contain, when full, just 1 lb., 4 ounces, and 1 ounce of powder, respectively;* a copper cylindrical tube of 3 feet, or 3 feet 6 inches long, by ¾ inch diameter from out to out; a set of three tubes of about an inch in diameter, 3 feet long each, with joints, so as to be screwed at pleasure into one length of 6 or 9 feet, or with more joints if deeper holes are employed, and so constructed that when together the interior should form one smooth surface: a copper funnel, the bowl large enough to contain about 1 lb. of powder, and the neck 2 inches long, by somewhat less than ¾ inch diameter; together with some coils of Bickford's patent fuse.

By means of the measures, the tubes, and the funnel, any

---

* Where the blasting is constant, and the charges vary but little, such as in sinking shafts, driving galleries, &c., it might be found convenient to have charges of the most usual quantities prepared previously in papers, cartridges, or perhaps even in little boxes or chargers.

specific charge of powder may be lodged *clear* to the bottom of any hole; and if horizontal, or nearly so, by pushing it in through the tubes with a wooden stick or ramrod.*

One end of a piece of the patent fuse is inserted well into the powder, the other end cut off about an inch beyond the outside of the hole; a little wadding is then pressed down over the powder with the tamping-bar, and upon that the tamping in the usual manner (but with a proper material) to the top, without any necessity for the moist clay.

Most of the accidents that occur to miners arise from the first blows of the tamping-bar over the charge; to obviate this, the first 2 or 3 inches of the tamping should be merely pressed down gently over the wadding, and then the hard ramming commenced over that: this cannot injure the effect of the explosion, as it is generally acknowledged that a small vacant space about the powder tends, if anything, to increase its power.

If the tamping-bar be tipped with brass, it will add more security, and at *very little* increased expense.

The outer end of the fuse is then lighted; there is neither difficulty, loss of time, nor extra cost or labour, by using these precautions, and obtaining all the consequent advantages; they should therefore never be neglected.

Whenever blasting is to be performed on an extensive scale within a limited space, it will be quite worth while to ascertain by a few experiments the value of the different ingredients that are to be used, or matter to be acted upon in the particular locality, as well as the best modes of applying them; such as the strength of the powder, the tenacity of the rock, the value of the tamping material, &c: it is clear that, by proportioning these justly, so as to obtain the greatest effect with the smallest means, much time and expense may be spared.

OF THE TRAIN AND FIRING.

The inconveniences and loss of time attending the ordinary mode of laying the trains for firing the charge in blasting holes have been mentioned above.

It is the very worst contrived part of the whole operation

---

* It affords one very important medium of security against accidents to deposit the powder quite clear to the bottom of the hole by means of these tubes, instead of allowing grains to hang on the sides, as they must do when it is poured in in the ordinary manner, particularly in holes that are inclined, where it is easy to conceive that a *regular train* may sometimes be left from top to bottom.

of blasting; but fortunately a most valuable improvement has been made of late years by the invention of Bickford's patent safety fuse.*

The use of this article is extremely simple; it is efficient in damp situations, and even under water, by using the quality prepared expressly for that object; a miss-fire is scarcely ever experienced, unless there be great carelessness, and it is a *very great* protection against accidents.

Whenever accidents have occurred (which are *extremely* rare), they have been traced to circumstances which could not be affected by the fuse; namely, to the first applications of the tamping-bar over the powder.

So large an opening for the escape of the powder is not created by the fuse-hole as by the needle;† and, taking everything into consideration, it is calculated that the use of the fuse is cheaper than the ordinary priming, even if the very trifling cost, or at least difference of either, could be deemed of importance.

SAFETY FUSE.

At Kingstown Harbour it was employed constantly in working with a diving bell in blasting rock for foundations in from 20 to 30 feet depth of water, and with complete success.

The following account of the great value of this invention is from a Paper by B. Mullins, Esq., an intelligent member of the firm who have the contract for the Kingstown Harbour works:—

"Rock-blasting operations have been for many years, and are now, carried on extensively by my firm. Bickford's safety fuse has been invariably used in those operations since

---

* This patent was granted on the 6th September, 1831, to W. Bickford, of Tucking Mill, Cornwall, for "an instrument for igniting gunpowder when used for blasting rocks, which he denominates the 'Miners' Safety Fuse.'" This fuse was therein described to be a cylinder of gunpowder or other explosive compound, inclosed within a hempen cord, which is first twisted and afterwards overlaid with another cord to strengthen the casing thus formed, then varnished to preserve the contents from injury by moisture, and finally covered with whiting, or other suitable matter, to prevent the varnish from adhering.—[ED.]

† In firing very small charges, the opening made by the needle or fuse-hole tends very much to reduce the effect. In the tamping experiments it will be observed that, of clay, and other compact tamping, portions of the top were frequently carried away; in all these and other partial removals of tamping it was observed that the principal openings were always on the side where the fuse had been.

the summer of 1833. It has our entire approval, as being more certain in its effects, less hazardous in its application, and ultimately cheaper, although not apparently so, than priming in the ordinary mode. From the period of its introduction up to the present time, we have not had an accident in any of our works from blasting, although within that period 73,600 lb. of powder have been consumed, and labour equal to that of 288,719 days of one man expended: nor had we more than two or three cases of miss-fire that I can recollect, and those were caused by want of caution in using improper tamping material, having stones in it which cut the fuse and severed the train. We have used of it in the interval 167,322 lineal feet.

"The cost of 167,322 feet of the fuse was 304*l*. 7*s*. 9*d*., while that of the actual quantity of powder required for the same amount of priming, namely, 35 barrels, would be 105*l*. besides the labour of the application of the latter."

Then follows a comparison between the old system of loading and firing, and the mode with the fuse.

After describing in detail the old practice (that, indeed, still used in most places), the statement continues thus:

"In the application of the fuse the charge may be lodged at any required depth in the rock. We lately drove a hole with a 5-inch gauge 20¼ feet horizontally into the face of the cliff at Dalkey, and charged it with 85 lb. of powder, by which we released 2000 tons of solid rock, a quantity far exceeding what could be displaced by vertical or oblique bores. Straw tubes could hardly be made of so great lengths, and, were it practicable, there is so great a loss of time, and so much uncertainty in using them in horizontal holes of much less depth, that they bear no comparison in facility of use to the fuse, which effectually supersedes the tedious and hazardous employment of the needle, and is a perfect preventive against premature explosion or miss-fire, wherein the old methods are particularly defective.

"It has this further advantage, that any number of shots may be simultaneously fired, whilst with the straw-tubes and match-paper not more than three, with a probable chance of escape to the men employed.

"In wet quarries the fuse is quite as effective as in dry: where wet joints are met with in boring, which frequently happens, the holes fill with water, and must necessarily, by the old methods, before being charged, be rendered perfectly dry; this is accomplished by wrapping coarse tow or mop thrums round the end of a stick, and mopping out the water; the finest argillaceous clay is then kneaded into a paste, and worked constantly upwards and downwards by the tamper struck with the hammer, and continued until the blaster is satisfied of having staunched the leaky joints in the hole: this being done, the hole is then cleared out with an iron scraper, made dry, and charged; and, after all this labour, success is doubtful. A man often spends half a day drying, tamping, and charging a hole in this way to no purpose, the powder having got wet.

"In these cases the fuse furnishes an effectual remedy; a waterproof bag containing the powder, and having a sufficient length of fuse closely tied in its mouth, is pushed home to the bottom of the hole, which is then tamped and fired with as much certainty and effect as in dry work.

"Blasting in deep water by means of the diving bell is rendered comparatively easy by the use of the fuse, as compared with the tedious, costly, and ineffective process heretofore practised.

"In the old way, the charge was lodged in a tin canister in the bored rock; in this canister a small tube was fixed, and raised joint after joint, until the bell was elevated above the water's surface; then a small piece of iron made red hot was thrown into the tube to fire the charge in the canister. What the effect produced may have been, I cannot say of my own knowledge, not having seen those operations; but I have been informed that, except in insulated rocks, it was of little use. The canister and the joint of the tube adjoining were blown to pieces, and most of the joints more remote were flattened by the collapse of the water, so that new canisters and tubes were necessary for every shot.

"The fuse employed in blasting under water is somewhat different, and more expensive than the other, and is called sump-fuse; it is used in the following manner:

"A waterproof charger or bag, containing the powder with a piece of fuse 5 or 6 feet long closely tied in its mouth, is dipped in boiling pitch * to secure the orifice from wet; the charge is then put into the bore-hole, which is tamped with sand, or the fine chippings of the stone-cutter's waste, and the fuse, the upper end of which is retained in the diving

---

* A *perfect* waterproof composition for this purpose is made of
8 parts by weight, pitch,
1 do. do. bees'-wax,
1 do. do. tallow,
} melted together, but not boiled.

bell, being set fire to, is thrust out under its edge into the water: the bell is then, by signal, removed 8 or 10 feet out of the way; the fuse burns through the water, and explosion follows; the proximity of the bell to the blasted rock, without endangering the workmen, enables them to resume their operations with little or no delay.

"In founding the Commercial Wharf wall (a considerable part on rock) in 22 feet water at low water of spring-tides and in clearing for abutment for setting frame for the eastern pier head in 28 feet at low water (the rise of tide 12 feet) we pursued the method above described with success.

"Having obtained a list of all the men who had been killed, or badly wounded, by the old methods of blasting at the quarries for the Kingstown Harbour works, previous to the use of the fuse, I annex it."

The list referred to of accidents under previous contracts, for the first 15 years, contains the names of 30 individuals, two of whom were twice injured, making 32 casualties, consequently more than 2 per year; it includes 7 killed, 4 loss of one eye, 1 loss of both eyes, and 20 injuries; while, during the 8 years of the present contract, there has been but one man injured, and that before the fuse was introduced.

In order to test the extent of applicability of this composition for blasting under water, pieces of the quality prepared for it (called sump-fuse), which is somewhat thicker than the ordinary kind, being about $\frac{3}{16}$ of an inch thick, were kept immersed (except the upper ends) for different periods up to upwards of 16 hours, and were then found to burn throughout with their ordinary force; no trial was made beyond that time.

Pieces 25 feet long (their usual dimension), having one end inserted into a few ounces of powder inclosed in a water-proof bag, were tied to long chains; the powder-bag was lowered into the water by a weight, and, the other end of the fuse being lighted from a boat, the whole was lowered again until the weight touched the bottom in 39 feet depth, and each time burned through and exploded the powder in 13 or 14 minutes. In order to get a greater length, an extra kind was procured in 51 feet and 52 feet lengths; it was apparently expressly made, and thicker than the other, being about $\frac{1}{4}$ of an inch in diameter.

This succeeded perfectly in every case; but, being provided for the purpose, it was not considered so satisfactory a trial, and appears to be unnecessary, considering the efficiency of that usually sold.

The common fuse (that not prepared for water) was tried, and on some occasions, when lighted very soon after being immersed, would burn through many feet under water, but frequently only a short distance; it is manifestly not adapted for water (and not assumed to be so), but it is perfectly efficient in damp situations, if fired without much delay.

There is one great merit in this fuse, namely, that the improvement is gained without the least sacrifice of simplicity; on the contrary, it is much easier to apply than any other process.

The only inconvenience attending its use, of which I am aware, is the length of time it takes to burn any but short pieces.

It burns at the rate of from 2 to 3 feet per minute; and as a minute usually affords ample time for the person firing to remove to a sufficient distance, there is a delay and an impatience created in watching the longer terms, particularly when it amounts to 3 or 4 minutes, or more.*

This lengthened time of burning would quite preclude its use for many military mines, where the explosion must take place at a particular instant.

In blasting under water, as it can only be lighted in the bell, or at the surface if the bell be not used in firing and the water deep, the time consumed in burning the fuse would be very inconvenient.

FIRING BY A VOLTAIC BATTERY.

Charges of powder have been fired under ground, as well as under water, and to considerable distances, by the voltaic battery.

It is to be presumed, however, that the distinct machinery for this purpose, the expense, and probably some degree of nicety in its arrangements, even after all the improvements that have been made at Chatham by Colonel Pasley, would render it inapplicable to ordinary purposes; although for firing *very large* quantities of powder, under very peculiar circumstances, it has been considered very useful. The tamping by this mode would be much more complete and substantial, having only two thin wires through it; and by the very instantaneous effect produced, even at varying distances, simultaneous explosions, that are impossible

---

* The fuse burns slower through a tamping of loose sand than through a tough material well rammed, but it is not extinguished by it.

by any other means, might be effected by this mode of ignition.*

MODE OF LIGHTING A TRAIN IN SHAFTS.

Under many circumstances the manner adopted for *lighting* a train is a matter of some consequence.

At the bottom of deep shafts, or of long small galleries where there is no shelter for the man who applies the light, the touch-paper, German tinder, or match of whatever kind, must afford time for him to get completely out of the way. It is attended with some difficulty to secure that object, without wasting time in doing so, by allowing too long an interval; but the latter must be the usual, as the only safe alternative.

The safety fuse has an advantage in this respect, as its period of burning can be more regularly calculated than that of any other match usually resorted to.

A French engineer has proposed and employed a manner of remedying this inconvenience, as far as regards blasting in a shaft, which would appear to be useful where the safety fuse is not employed.

One end of a wire is fixed temporarily into a little powder connected with the train, the other end in a coil being at the surface of the ground.

The wire is carefully straightened up the shaft from any sharp twists or bends; a piece of German tinder is passed round it, forming a very loose ring, and, being lighted, is let go, and drops down on the powder.

The wire may of course be constantly suspended to the side of the shaft, and fixed and applied from time to time as wanted, being lengthened from the coil above as occasion requires.

To facilitate the descent, if necessary, a small weight might be easily added, to the under part of which the lighted match could be fixed.

MEANS FOR PREVENTING SMALL STONES FLYING ABOUT IN BLASTING.

A great loss of time and labour is experienced in quarries and other confined situations, arising from the necessity for

---

* In the Appendix is given a brief account of the successful application of the voltaic firing in removing the Round Down Cliff, at Dover for the works of the South-Eastern Railway.—[ED.]

the workmen to retire to a distance at every blast; not only those engaged in the precise operation, but all others who can be at all supposed to be within reach of its effects.

This is a greater evil than what is perhaps commonly thought.

I have myself been in a great quarry, and in the course of half an hour seen 20 or 30 men, at given signals, retire two or three times from five or six different parts at which they were at work, to which they went leisurely back again after each explosion or ascertained miss-fire.

There is another occasion where the necessary retiring of the miner from the effect of a blast is attended with peculiar inconvenience; that is, from the bottom of a shaft, in which case he has to be drawn up to the top.

When a better mode of applying the charge shall be generally understood, there will be fewer occasions of the projection of stones, but there must always be some; any remedy for this waste of time, therefore, must be very valuable.

QUARRY SHIELDS.

In some quarries in the neighbourhood of Glasgow they are in the habit of frequently applying a piece of old boiler iron, of about 2 feet 6 inches square and $\frac{1}{4}$ inch thick, over the hole when fired, which acts as a shield, and, in small blasts, so far prevents any flying about of loose stones, that the men take much less precaution in moving out of the way on those occasions than when it is not used.

As they have horizontal as well as vertical faces to work on, the shield is suspended over the horizontal holes, and laid flat over those that are vertical; in the latter case a large stone or two, if at hand, is frequently placed upon it.

This application might be very useful under particular circumstances of blasting; for instance, where the blasts are generally small, and in confined situations. In a shaft, where the holes will be all probably vertical, or nearly so, and the blasts not large, a good shield could be placed over every hole, and weighted either with stones, or with one or two half-hundred weights, kept in the shaft for the purpose. On trial at the particular place, this might be found to give such *certain* security, that the miners would not require to be removed more than a very short way above, either by the bucket, or by a ladder, which would lead to a very great saving of time, labour, and expense.

ON TAMPING.

The desideratum in tamping is to obtain the greatest possible resistance over the charge of powder; if it could be made as strong as the rock itself it would be perfection.

If 12 inches of one species of tamping will afford as much resistance as 18 inches of another, the question will be—Does the former, in the application, require as much more time and expense as the operation of boring the additional 6 inches of hole? If not, it will, under most circumstances, be better in the relative proportion of that expenditure.

Where other qualities are equal, that which can be applied in the least space of time will be far preferable, particularly in such operations as sinking shafts, driving long narrow galleries or driftways, and other situations where the progress is necessarily slow.

Different materials are employed for tamping:—

I. The chips and dust of the quarry itself. This is what is most commonly used, unless there be flint or other stone in it that notoriously strikes fire.

II. Sand poured in loose, or stirred up as it is poured in, to make it more compact. This is an approved material in many places, and is recommended to be very fine and dry.

III. Clay, well dried, either by exposure to the sun, or what is more certain and more rapid, by a fire.

Wherever blasting is going on, there must be smiths' forges at work: the clay is formed by the hand into rolls of about 2 inches in diameter, and readily dried by the smiths' fires.

In the course of some experiments, clay was used that was in a state of powder, owing to its having been dried a long time previous, and was not thought so good as when applied in a state just caked enough to remain in lump.

IV. Broken brick is an approved material in some localities, as being less liable to accidents by striking fire than chips of stone.

It is used in small pieces and dust, and is improved by being slightly moistened with water during the ramming.

Vegetable earth, or any small rubbish, is sometimes applied instead of the stone chips, when the latter are considered dangerous. Such are the simple ingredients used in tamping; that is exclusive of the addition of any mechanical contrivance: of these, the most essential to analyse is the application of sand, since its use has been by many strongly recommended, and, if efficacious, would be most convenient, on account of the rapidity and security with which it can be applied.

In Cachin's Mémoire sur la Digue de Cherbourg, printed in 1820, (a most interesting work as regards the construction of breakwaters,) it is thus stated in a note:—

"In blasting rock at Cherbourg, the use of the needle, and well-rammed tamping, has been long abandoned.

"The priming is in the usual manner, by straws, and the tamping is of very fine dry sand, poured in.

"It has been proved by long experience that the effect of the explosion is as great by this method as by the more laborious tamping in the usual manner."

And in the Journal of the Franklin Institute (United States) for July and August, 1836, after quoting a variety of experiments on the resistance of sand to motion through tubes, made as well in France as in America, and facts regarding the bursting of musket barrels, &c., by charges of sand over the powder, the conclusion come to is, that,

"Experience proves that the resistance offered by sand is quite sufficient for blasting rocks, and it is less troublesome, and more safe, than the usual mode."

It is added, that,

"To ensure success, the space left above the powder should have a length of ten or twelve times as great as the diameter of the hole."

General Pasley, on the contrary, asserts that sand as an ingredient for tamping was found at Chatham to be utterly valueless; but acknowledges that the opportunities there of blasting were few and on a very small scale. Many other officers of the British Engineers have been long under the same impression.

It does not appear that any of these opinions have been formed upon any more precise experiments than the sensible effect upon stones or rock with the usual charges; and as these effects are very different under circumstances that are apparently similar, and as they might vary with different proportions of the powder and ingredients used, advantage was taken of the opportunity afforded by the works carrying on at Kingstown Harbour, near Dublin, to try a few experiments that should be somewhat more definite.

The principle which it was thought would be most conclusive was, not to form a judgment by the effect produced on the rock, but to endeavour, if possible, to obtain a charge of powder that should in each case just blow out, or sensibly affect, the *tamping;* and a comparison of the charges to produce that effect would afford a strong proof of the relative value of the different systems ; thus—

If it should be shown (as will be found in the following table of experiments) that ¼ of an ounce of powder would blow out the sand tamping which filled a hole of 1 inch diameter and 2 feet deep, while 3 ounces in a similar hole would not disturb a well-rammed clay tamping, it afforded a perfect confirmation of General Pasley's statement, that the sand was good for nothing, as far as the use of it on that scale went.

The experiments detailed in the following Tables were made in granite rock, and, as far as could be judged, where it was firm and without fissures.

The charges were of ordinary merchants' blasting powder, procured from the contractors.

The sand was sharp or gritty, quite dry, and simply poured in over the powder, with a little wad of hay intervening. It was from the sea-shore, and of different qualities.

The finest was a clean running sand, fine enough nearly for an hour-glass, and weighed . . . . 65 lb. per cube foot.

The second quality, a middling gritty sand, and weighed . . 93 lb. „

The third, very coarse, or rather very fine sea shingle, the particles being from the size of a pin's head to that of a pea, and weighed . . 98 lb. „

The clay had been dried at the fire of a smith's forge, and was well rammed down in the usual manner.

Experiments to try the Comparative Value of Sand and Clay for Tamping.

F Sand, fine.   M Sand, middling.   C Sand, coarse.

No. of experiments.	Depth of hole.	Diameter of hole.	Charge of powder.	Description of tamping	Effect and Remarks.
No.	feet.	inches.	lb. oz.		
1	2	1	0 2	Clay.	$\frac{1}{2}$ inch of top of tamping removed—rock fractured.
2	2	1	0 2	do.	Tamping remained—rock star-fractured.
3	2	1	0 2	do.	8$\frac{1}{2}$ inches of tamping removed—rock fractured.
4	2	1	0 2	do.	Tamping remained—rock fractured.
5	2	1	0 3	do.	Portions of tamping adhered to sides of the hole—a mass of rock blown off.
6	2	1	0 3	do.	Tamping all remained—rock fractured.
7	2	1	0 3	do.	1$\frac{3}{8}$ inch of tamping removed—rock fractured.
8	2	1	0 3	do.	Tamping remained entire—rock fractured.
9	2	1	0 4	do.	5$\frac{1}{2}$ inches of tamping removed—rock fractured.
10	2	1	0 4	do.	2 inches of tamping removed—rock star-fractured.
11	2	1	0 2	Sand.	Tamping entirely blown out—rock uninjured.
12	2	1	0 1	do.	do. do.
13	2	1	0 1	do.	do. do.
14	2	1	0 0$\frac{3}{4}$	do.	do. do.
15	2	1	0 0$\frac{1}{2}$	do.	do. do.
16	2	1	0 0$\frac{1}{4}$	do.	do. do. This was repeated three times with the same effect. N.B.—By the ordinary miners' rule of allowing $\frac{1}{3}$ depth of hole for the charge, one of 2 feet deep by 1 inch diameter would require 3$\frac{1}{2}$ ounces.
17	4	1$\frac{1}{2}$	0 6	Clay.	1 foot 8-inches of top of tamping removed—rock fractured.
18	4	1$\frac{1}{2}$	0 6	Sand.	Tamping blown clean out—no fracture. N.B.—In hole 4-feet by 1$\frac{1}{2}$ inch, $\frac{1}{3}$ of depth will contain 15 oz. of powder.
19	6	2	0 8	M Sand	Tamping blown out, no fracture.

No. of experiments.	Depth of hole.	Diameter of hole.	Charge of powder.	Description of tamping	Effect and Remarks.
No.	feet.	inches.	lb. oz.		
20	6	2	0 8	C Sand.	4 ft. 7 inches of tamping blown out, below which was a hard crust of about 4 inches thick, and in the cavity below there remained 8 or 9 inches of the sand, mixed with burnt powder—no fracture in the rock.
21	6	2	0 10	Clay.	Tamping unmoved—rock uninjured.
22	6	2	0 10	Sand.	1 inch of top of sand removed—rock fractured.
23	6	2	0 10	M do.	Tamping blown out—no fracture in rock.
24	6	2	0 12	Clay.	2 inches of tamping removed—rock fractured.
25	6	2	0 12	Sand.	2 ft. 4 inches of sand removed.
26	6	2	0 12	M do.	Tamping all blown out—no fracture.
27	6	2	0 14	Clay.	1 in. removed—rock fractured.
28	6	2	0 14	Sand.	7 inches removed—rock fractured.
29	6	2	1 0	Clay.	Tamping unmoved—rock fractured thoroughly.
30	6	2	1 0	Sand.	Tamping blown out—rock fractured thoroughly.
31	6	2	1 0	C do.	Tamping unmoved—rock lifted—the explosion probably escaped through the joints.
32	6	2	1 4	C do.	Tamping blown out—rock star-fractured.
33	6	2	1 8	C do.	Blown out, rock slightly cracked.
34	6	2	1 8	C do.	Tamping blown out—rock shaken, joints opened.
35	6	2	1 8	C do.	Tamping blown out—no fracture.
36	6	2	1 8	C do.	4 feet 1 inch blown out, then 3 inches of loose sand, and below that a hard crust—no fracture of rock.
37	6	2	1 8	Clay.	Tamping unmoved — rock cracked.
					N.B.—In hole 6 ft. by 2 inches, ⅓ of depth will contain 2 lb. 8 ounces of powder.
38	9	2	0 6	M Sand	7 feet 10 inches blown out—no fracture in rock.
39	9	2	0 8	M do.	8 feet 6¼ inches do. do.
40	9	2	0 8	C do.	7 feet 4 inches do. do.
41	9	2	0 10	C do.	7 feet 1¼ inch. do. do.

No. of experiments.	Depth of hole.	Diameter of hole.	Charge of powder.	Description of tamping	Effect and Remarks.
No.	feet.	inches.	lb. oz.		
42	9	2	0 12	M Sand	Tamping blown out, no fracture.
43	9	2	0 12	C do.	7 ft. blown out—no fracture—a hard substance 1 ft. 3 in. above bottom of hole; then a cavity of 2 or 3 inches, and under it sand mixed with burnt powder; the crust required considerable labour to pierce with a 2-inch churn jumper, worked by 2 men.
44	9	2	0 14	C do.	6 feet 10½ inches blown out—no fracture—similar hard crust to preceding No.
45	9	2	3 0	Clay.	Tamping unmoved—rock very slightly cracked. N.B.—⅓ of hole 9 ft. by 2 in. will contain 3¾ lb. of powder.
46	9	2½	1 0	C Sand	3 ft. 11 in. blown out—no fracture—similar hard crust found.
47	9	2½	1 0	M do.	Blown out, except a small crust of about ¼ inch thick.
48	9	2½	1 8	C do.	5 feet 3 inches blown out—rock cracked.
49	9	2½	2 0	C do.	Tamping unmoved.
50	9	2½	3 0	C do.	Tamping unmoved—the explosion found vent by a side joint near the charge—making an extensive fracture.
51	9	2½	3 0	Clay.	Tamping unmoved—rock fractd.
52	9	2½	3 8	C Sand	5 feet 10 inches blown out—no fracture.
53	9	2½	4 0	Clay.	Tamping unmoved—rock fractd.
54	9	2½	4 0	C Sand	3 feet 2 inches blown out—no fracture—the hard crust 5 ft. 7 inches from top of hole was some inches thick.
55	9	2½	4 8	C Sand	Blown all out—rock slightly fractured.
56	9	2½	5 0	C do.	Blown all out—rock fractured.
57	9	2½	5 0	Clay.	Tamping unmoved—rock fractured.
58	9	2½	6 0	do.	Tamping unmoved — rock shaken — explosion escaped through side joints. N.B.—⅓ of depth of hole 9 ft. by 2½ inches would contain 5 lb. 14 oz., or nearly 6 lb. of powder.

In the same paper of the Franklin Institute above referred to, it is stated that the sand was sometimes poured in loose, and sometimes carefully *packed*.

The packing was performed by means of a small sharp stick, which was worked up and down as the sand was slowly poured in.

It is stated, that—

"This method was found to be the best, and is the one always used at Fort Adams, in charging drill holes for sand blasting.

"Sand that was packed presented a much greater resistance than that which was poured in loose."

In order to try the value of thus packing the sand, it was ascertained how much was added to the mass by this mode of condensation; for this purpose we used a tin tube of 24 inches long by $2\frac{1}{8}$ inches diameter at mouth.

$2\frac{1}{16}$ do. do. at bottom.

When filled with the different qualities of sand, the weights were respectively—

	Fine oz.	Middling oz.	Coarse. oz.
Poured in loose	65	76	80
Packed by process above described	74	$83\frac{1}{2}$	84

Many trials in blasting were made with tamping of sand thus packed.

In holes 2 feet deep by 1 inch diameter, each of the three qualities of sand was tried with a $\frac{1}{2}$-ounce and a $\frac{1}{4}$-ounce charge, and in the whole of them the sand was all blown out.

QUARRYING AND BLASTING ROCKS.

Experiments on Tamping with Packed Sand.

F Sand, fine.   M Sand, middling.   C Sand, coarse.

No. of experiments.	Depth of hole.	Diameter of hole.	Charge of powder.		Description of tamping	Effect and Remarks.
No.	feet.	inches.	lb.	oz.		
59	4	1½	0	7	C Sand	Tamping blown out.
60	4	1½	0	7	C do.	Tamping remained—rock fractured.
61	4	1½	0	6	M do.	2 feet 6¼ inches removed—rock fractured.
62	4	1½	0	6	M do.	1 foot 5¼ inches removed—rock fractured.
63	4	1½	0	6	C do.	Tamping blown out—but a slight crust adhered to the sides of the hole.
64	4	1½	0	5	M do.	Tamping blown out.
65	4	1½	0	5	C do.	Tamping removed—except a hard crust about 12 inches above bottom of hole.
66	4	1½	0	5	C do.	do. do.
67	4	1½	0	4	M do.	3 in. of tamping only remained.
68	4	1½	0	4	F do.	Tamping blown quite out.
69	6	2	0	7	F do.	
70	—	—	0	6	—	In all these, the tamping blown out.
71	—	—	0	5	—	
72	—	—	0	4	—	
73	—	—	0	3	—	All blown out but 1½ inch.
74	—	—	0	12	M do.	
75	—	—	0	11	—	
76	—	—	0	10	—	In all these, the tamping blown out.
77	—	—	0	9	—	
78	—	—	0	8	—	
79	—	—	0	7	—	
80	—	—	0	10	M do.	About 1 foot of sand remained rock fractured.
81	—	—	1	0	C do.	Tried twice—in both cases the tamping blown out.
82	—	—	0	15	—	Blown out, except a hard crust 17¼ inches from bottom.
83	—	—	0	15	—	2 feet removed—a hard crust, took 2 men 40 minutes to bore through.
84	—	—	0	14	—	16 inches remained — hard crust formed.
85	—	—	0	13	—	Blown out, all but hard crust.
86	—	—	0	13	—	3 feet 6½ inches blown out—hard crust 12 or 14 inches thick, bored through with much labour.
87	—	—	0	12	—	5 feet 4½ inches blown out.

The quality of the sand has not been always noted in these or the former trials; but all three kinds were employed. The very fine, contrary to the received opinion that it is the best, universally failed: the very coarse, in the larger explosions, was, for a few inches in thickness, generally vitrified or cemented together into a very hard crust. On trial with an acid, it was found that there were some particles of limestone mixed with this sand, the sudden burning of which might perhaps have assisted in producing this effect.

In all the descriptions of sand, indeed, this cementing process took place at times, more or less, and all of them contained particles of lime.

These last experiments were made at a different period from the former trials of sand, and probably with some variety in the quality of the powder and material, which must account for the apparent inferiority of the packed sand; whereas the *packed* must be the best, although in too small a degree to remedy the inherent defects in any tamping of sand.

In fact, it is impossible to reduce these kinds of experiments to any very close results *in detail*, although by numerous trials we may come to *general results* that may be well relied on.

The conclusion come to from all these experiments, notwithstanding some discrepancies that will be observed, is, that sand of any description, and however applied, is, when used by itself for small blasts, perfectly worthless, and quite inferior to clay tamping for larger explosions, at least so far as for holes 9 feet deep by $2\frac{1}{2}$ inches in diameter.

The cause of the sand tamping not presenting the same resistance to the explosion of powder in a blasting hole that it does to the mechanical pressure through tubes, is, no doubt, that the explosion penetrates among the particles, and loosens and separates them, instead of wedging them together as when pressed. The same action will be observed in a subsequent part to produce an extraordinary effect on sand or other loose material when used with iron cones.

No examination was made in any of these or preceding experiments of the quality of the powder, excepting those expressly for that object. It was in all cases the ordinary merchants' blasting powder; nor was any account taken of weather, or other such circumstances as might cause discrepancies in trials made at different periods.

The only other materials for tamping, requiring much notice, are—

> Broken brick, and
> Broken stone, or quarry rubbish.

I. *Broken brick* is a good material for tamping, and gives considerable resistance, though not so much as clay; *vide* Nos. 116, 117, 118, in Table. It will not strike fire with iron, and it is to be procured in most situations.

II. *Broken stone* is of two qualities. Some quarrymen are in the habit of using a rotten kind of stone that is found in most stone countries; not being brittle, it rams into a very firm mass, and is not subject to strike fire; but being *stone* at all, gives an opportunity for careless workmen to apply the harder quality instead, or, at any rate, to mix pieces of the latter with it, by which the operation is subject to the contingencies of the use of the hard material.

The ordinary material used for tamping is the broken stone and rubbish of the quarry itself, unless notoriously subject to strike fire.

This, by experiment with the blue limestone or granite of the neighbourhood of Dublin, was found to be inferior to clay as a resisting medium. *Vide* experiment No. 119.

It is also more liable to cut the safety fuse, or to derange the needle hole, but, above all, it is never, in any rock, entirely free from some danger of giving fire, and causing accidents: this is quite enough to occasion it to be rejected wherever it is possible to procure a substitute.

Experiments with Tamping of Clay, Broken Brick, and Broken Stone.

C Contractor's Powder.   G Government Powder.

No. of experiments.	Holes. Depth.	Holes. Diameter	Charge and description of powder.	Tamping.	Results and Observations.
No.	ft. in.	in.			
88	3 0	2	2 oz. C.	Clay 4 to 16 in.	All blown out.
89	3 0	2	,,	Do. 17 in.	A small quantity left.
90	3 0	2	,,	Do. 18 in.	All blown out.
91	2 0	2	,,	Do. 19 in.	7 inches of tamping removed.
92	2 0	2	2 oz. G.	Do. 19 in.	7½ inches removed.
93	3 0	2	2 oz. C.	Do. 20 in.	5¼ inches removed.
94	2 0	2	2 oz. G.	Do. 20 in.	8¼ inches removed.
95	3 0	2	2 oz. C.	Do. 24 in.	3 inches removed.
96	2 2	2	2 oz. G.	Do. 24 in.	6 inches removed.
97	2 10	2	3 oz. C.	Do. 32¾ in.	Tamping undisturbed — rock well cracked—line of least resistance 19 inches in a favourable position, near a salient angle.
98	3 2	2	3 oz. G.	Do. 37¾ in.	Tamping blown out—rock, to all appearance, not affected—line of least resistance 2 feet, in an unfavourable position, near a re-entering angle.*
99	4 1½	2	2½ oz. C.	Do. 4 ft.	16 inches removed — the clay in this case was very dry, and in powder, and considered to be less efficient than when caked.
100	3 0	2	4 oz. C.	Do. 24 in.	Tamping not disturbed — rock lifted.
101	1 7	2	1 oz. C.	Do. 10 in.	All blown out.
102	1 2½	2	1 oz. C.	Do. 12 in.	About 3 inches of top blown out, and a considerable portion round fuse-hole.
103 to 105	3 0	1	2 oz. C.	Do. 5 & 6 in.	All blown out.
106	3 0	1	2 oz. C.	Do. 7 in.	1½ inch removed.
107	3 0	1	2 oz. G.	Do. 7 in.	All blown out—tried twice.
108	2 9½	1	2 oz. G.	Do. 8 in.	¾ inch removed—rock slightly affected.
109	3 1	1	2 oz. G.	Do. 8 in.	2½ inches removed.
110	4 0	3	2 oz. C.	Do. 14 in.	All blown out.
111	3 9½	3	2 oz. C.	Do. 18 in.	All blown out.
112	2 0	3	2 oz. C.	Do. 20 in.	Very little removed, except on side of fuse-hole.
113	3 10½	3	2 oz. C.	Do. 22 in.	13 inches removed.

* The effect on the tamping in this experiment (No. 98) is so different from all the others, with even superior charges, (*vide* Nos. 100, and 114 to 122,) that there was probably some misapprehension in preparing or recording the experiment.

No. of experiments.	Holes.		Charge and description of powder.	Tamping.	Results and Observations.
	Depth.	Diameter			
No.	ft. in.	in.			
114	4 0¼	3	4 oz. C.	Clay 26 in.	17½ inches removed.
115	4 0¼	3	4 oz. C.	Do. 30. in.	22 inches removed.
116	3 ·2	2	4 oz. C.	Do. 24 in.	Tamping undisturbed — rock slightly cracked.
117	3 2	2	4 oz. C.	Do. 22 in.	3 inches of top of tamping removed—the remainder undisturbed.
118	3 2	2	4 oz. C.	Do. 21 in.	4 inches of tamping removed—rock burst.
119	3 2	2	4 oz. C.	Do. 20 in.	1 inch removed—rock burst.
120	2 0	2	4 oz. C.	Do. 18 in.	6 inches removed—rock burst.
121	2 0	3	4 oz. C.	Do. 20 in.	Rock burst—enlarged upper part of fuse-hole.
122	2 0	3	4 oz. C.	Do. 18 in.	8 inches removed—rock burst.
123	16 0	4	35 lb. C.	Do. 9 ft.	The charge occupied 7 feet—the rock was separated across the hole, at the outer end of which 4 inches of the tamping was found firmly adhering to the side of the half hole in the solid rock.
124	7 2	2	2 lb. 8 oz. C.	Do. 5 ft. 2 in.	The powder occupied 2 feet—the rock was separated so as to cut the hole longitudinally in two—the tamping was found adhering firmly to 15 inches of the upper end of the half hole.
125	11 0	3	19 lb. 12 oz. C.	Do. 4 ft.	The powder occupied 7 feet out 11—the face of the rock was blown down along the line of the hole, excepting the upper 2 feet 6 inches, which remained, and in which the tamping continued firmly fixed after the explosion. Fig. 13, p. 49.
126	2 0	2	2 oz. C.	Broken brick, quite dry, 18 in.	All blown out.
127	2 0	2	2 oz. C.	Do. 23 in.	All blown out.
128	2 0	2	2 oz. G.	Broken brick, damped, 23 in.	All blown out, except a slight incrustation on sides near the bottom.
129	2 0	2	2 oz. C.	Broken granite stone, 23 in.	All blown out—the stone was nearly in a state of disintegration.

But few trials were made with broken brick; the object was merely to confirm the opinion that it possessed no decided advantage, in point of resistance, over clay; if anything, it is believed to be somewhat inferior, and requires a little moisture, which is a slight disadvantage; and it is more liable to have particles of stone, or hard material that might strike fire, mixed up with it.

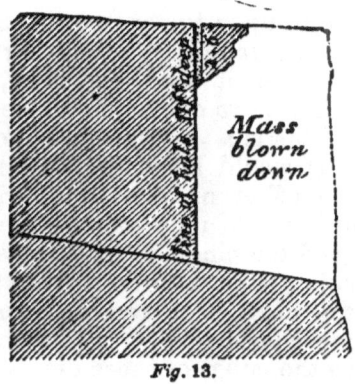

Fig. 13.

The broken stone was tried once or twice besides the instance recorded, and in all showed an inferiority to the clay; the use of it being always attended with more or less of danger, it was not thought advisable even to *experiment* much with it.

It will be perceived that in holes of 2 inches diameter, 2 ounces of powder will blow out about 18 inches of clay, and not more.

In holes of 1 inch diameter, 2 ounces will not blow out above 7 inches.

In holes of 3 inches diameter, 2 ounces will not blow out above 19 or 20 inches.

This comparison, however, is not quite conclusive enough to found a theory on, as the position of so small a quantity as 2 ounces spread on the surface of a 3-inch hole gives it a disadvantage, whereas in a 1-inch hole it lies very compact.

Increase of charges does not produce the increased effect upon good tamping that might be expected. It has been shown (Nos. 90, 91, Table of Experiments,) that in a 2-inch hole 2 ounces of powder will just blow out 18 inches of the tamping; in No. 100, and from 116 to 122, it will be found that 4 ounces, that is, double the charge, had scarcely, if at all, more effect, so far as can be judged under the different circumstances.

When, however, the rock is opened by the explosion, the effect on a tamping of clay or other tough material is greatly reduced; see Nos. 123 and 124, and particularly No. 125, for a remarkable instance of this; also Nos. 97 and 98, for the difference in effect on the tamping caused by the rock yielding, or not, to the explosion. It would appear that the

D

action upon the rock in opening it is much more rapid than on the tamping; even where the rock is separated across the line of the hole itself, the tamping is usually found adhering to one or both sides.

This is a very favourable circumstance in blasting.

It would be interesting to follow up the experiments of the effect of varying charges, with varying depths and diameters, upon one uniform description of clay or other good tamping; it would seem probable that after a certain point, which may be perhaps about 24 inches in a 2-feet hole, the charges to remove increasing depths of such tamping must be increased in a much greater proportion even than as the cubes of those depths: it is very difficult, however, to make such experiments, on account of the bursting of the rock with increasing charges, by which the effect on the tamping is reduced; it could only be done by holes in re-entering angles, *very closely* bound by projecting masses.

After having tried the value of the ordinary modes of tamping with broken stone, sand, brick, and clay, it becomes worthy of consideration whether additional resistance might not be obtained by some mechanical application of a different nature, tending to save time, labour, and chances of accident.

Any such contrivance, to be practically of general service, must be very simple in construction and application, and obtained at an expense not disproportionate to the advantages gained by its use.

The one that naturally suggests itself is some kind of plug or wedge, fixed in the loaded hole in a manner to increase the resistance.

If such a plug could be contrived to give, with sand, or loose small broken stone, or quarry rubbish, equal resistance to the same depth of good clay tamping, a very great advantage would be obtained in rapidity* of the tamping, and in security from accidents.

Many trials were made for this purpose.

The first was with an iron plug two or three inches long, and very slightly coned, the larger end being of somewhat greater diameter than the hole.

It is well known that such plugs, when driven into a hole in rock, and not having perhaps half an inch of contact, will raise from the ground the weight of many tons, showing a

---

* Substantial tamping is usually executed at the rate of from six to twelve inches per minute; nine inches per minute may therefore be calculated upon as a medium for holes of almost any size.

degree of tenacity that it was expected would be very powerful against the explosion from within; but when tried over the sand, although evidently affording much increased opposition, they were still in all cases driven out with sharp explosions, by smaller quantities of powder than would have removed good clay tamping.

Plugs of wood were tried with a similar result.

Iron plugs, of a form *slightly* curved in a barrel shape, and about 3 inches long, were also tried, and with better effect: when tightly driven into the top of the hole they took firm hold, and over loose sand, when properly fitted, appeared to give more resistance than equal depths of clay (*vide* Nos. 130, 131, 132 of Tables); they had a groove along the side for the fuse to pass through (fig. 14), and a strong eye to which

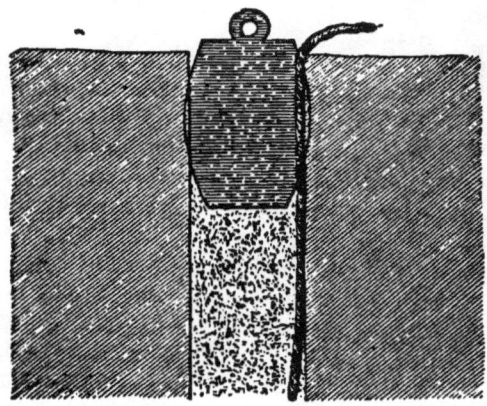

Fig. 14.

a string could be fastened with some object attached, to enable it to be seen and found if the plug was forced into the air.

A few of them were made for 2-inch holes, that is, varying in diameter from a little less to a little more than 2 inches; so that from the set it was easy by trial to find one that would fit with a proper degree of tightness.

The objections to the use of such plugs would be—expense; occasional losses; when blown out, some danger of falling on the by-standers; a degree of difficulty in removing them when not blown out, or the rock cracked precisely across the mouth of the hole; and want of simplicity by requiring an additional implement;—in this case, too, there is the chance of the workmen carelessly applying such as would not fit well, by which they would be rendered of little or no service.

Pins formed of cylindrical pieces of iron, 6 inches long, whereof two sides were taken off, each of them of the thickness at top of about a quarter of the diameter of the cylinder, and tapering to nothing at the bottom, so as to form two wedges, were then tried over sand at the upper edge of the hole; and although the pin was driven up against these wedges, and much effect produced, still they were not equal to the clay.

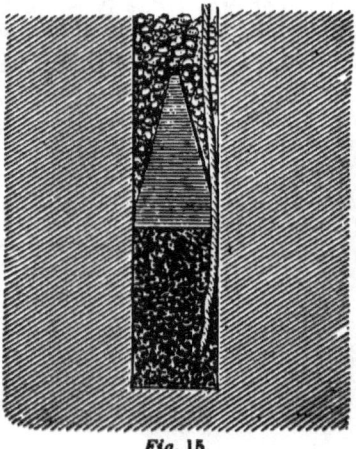

Fig. 15.

A cone placed immediately on the charge of powder, and filled over with stone broken to pieces of about half an inch cube (fig. 15), it was considered would give very powerful effects. Such a cone at the end of a rod of iron, and having only 16 inches of broken stone over it, at Chatham, where the idea originated, was proved to support a weight of 16 tons. A similar trial was made subsequently at Kingstown, and it supported a weight of 10 tons without showing any signs of yielding, and even with fine sand, instead of stone, over it; and although the base of the cone was only $1\frac{3}{4}$ inch in a hole of $2\frac{3}{8}$ inches in diameter, it sustained the same weight perfectly.

Still, notwithstanding this great power, when opposed to the gradual application of a force from above, it was found to give way to the explosion of the powder from below (see Nos. 133 to 145).

It would seem that the explosion penetrates round the sides of the cones, however small the windage (and they cannot be made to fit very tight, on account of the irregular size and shape of the holes), and by the fuse hole, and acts directly on the broken stone or sand above, so as to prevent them from operating as a wedge.

Some of the objections to the barrel-shaped plugs apply to these cones also: it would be attended with great labour to extract them if not disengaged by the explosion; and it would be very difficult to apply them in holes horizoutal, or nearly so, unless they were very shallow.

The last contrivance tried, and which, as far as it has gone, has given a greater degree of resistance than any other mode of tamping, has been the application of iron cones, with long iron arrows applied as wedges (fig. 16).

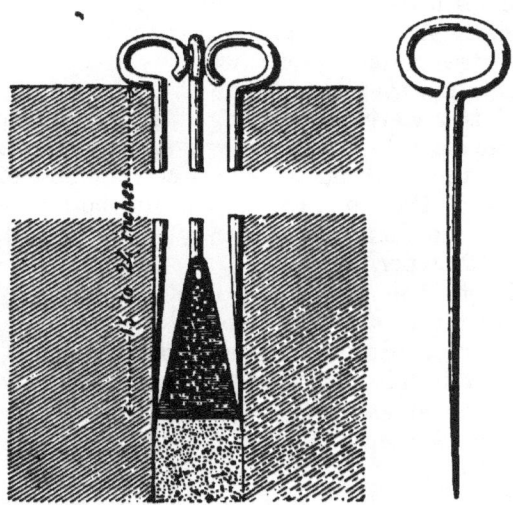

Fig. 16.

The cones may be from 3 to 6 inches long, with a groove for the fuse, and an eye at the top.

The arrows, from 21 inches to 2 feet 6 inches long, made of $\frac{1}{2}$-inch round iron, with a long fine point at one end, and turned to a handle at the other.

The arrows are proposed to be of this shape, as perhaps the cheapest that could be made.

These cones were inserted at from 6 inches to 2 feet below the edge of the hole: for their effects, see Nos. 146 to 163 of the following Tables.

The arrows should not be less than $\frac{3}{8}$ inch thick (see No. 152), nor more than $\frac{4}{8}$; nor fewer in number than 3 for a 2-inch hole, or 5 for a 3-inch (see Nos. 152, and 161, 162).

The cone is let down over the sand, clay, or rubbish, and the arrows are fixed to their position by a slight blow or two with a hammer or stone.

The objections to this implement are in some respects

similar to the others; namely, expense, chance of occasional losses when blown out, or of falling on people's heads; the almost impossibility, frequently, of removing them when not blown out, without breaking open the rock; and want of simplicity in requiring so many additional implements.

The wear and tear of the arrows would be considerable, as they are usually thoroughly flattened near their lower end, and sometimes broken by the explosion.

One remarkable circumstance occurred repeatedly in the trials with these cones and arrows.

With sand laid over the charge of powder, and *under* the cones, the latter were firmly wedged, while the sand (sometimes in considerable quantity) was blown out entirely, or nearly so, by the small opening between the base of the cone and the side of the hole, the cones not being particularly loose, or more so than was necessary to go down freely (see Nos. 146, 149, experiments).

Even clay to the depth of 6 inches was removed in the same manner from *under* the cone (No. 156).

Broken stone could not escape in a similar way, but the explosions through it were observed also to find a vent round the sides of the cones.

From 8 to 12 inches of clay tamping over the sand prevented this effect.

In all cases the powder was poured in by a copper tube to the bottom of the hole, and a very thin covering of one or two folds of paper was the only wadding used.

The rock was not affected, so far as could be perceived, unless where otherwise mentioned.

Clay, and broken brick tamping, were in all cases firmly rammed down.

Sand and gravel in all cases poured in loose.

Broken stone, when employed by itself, was well rammed; and when used with any kind of plugs or cones, it was poured in loose.

Where *portions* of the tamping were removed, it was considered to have been occasioned by the escape of the explosion up the fuse hole, or round the cones, and not by the general concussion.

Experiments on Tamping with different kinds of Iron Plugs or Cones.

No. of experiments.	Holes. Depth.	Holes. Diameter.	Charge and description of powder.	Description of tamping.	Results and Observations.
No.	ft. in.	in.			
130	2 0	2	2 oz. Contr.	20 inches of sand, then an iron plug of barrel shape, 3 inches long, driven in very firmly.	Plug not stirred.
131	6 0	2	1 lb. 8 oz. Contr.	Fine sand over charge to within 3 in. of top, then barrel-shaped plug.	Plug unmoved—rock cracked—3 feet 4 inches of sand remained in hole.
132	6 0	2	1 lb. 12 oz. Contr.	Tamping as in preceding trial.	Plug and tamping unmoved—rock thoroughly rent.
133	2 0	2	1 oz. Contr.	Iron cone 6 in. long by 1⅞ in. diameter at base, over charge, then sand to top.	All blown out.
134	6 0	2	2 oz. Govt.	Iron cone 6 inches, then fine sand to top.	4 feet 5 inches of sand removed, cone and 1 foot of sand remained.
135	3 0	2	2 oz. Contr.	4 inches fine sand, then iron cone, then sand to top.	All blown out.
136	5 1½	2	2 oz. Govt.	2 inches of sand, then iron cone 2½ inches long, with sand to top.	Blew all the sand out—the cone remained at the bottom of the hole.
137	2 0	2	2 oz. Govt.	Iron cone 3¾ in. then gravel to top.	All blown out.
138	6 0	2	2 oz. Govt.	Iron cone 6 inches, then gravel to top.	2 feet 9 inches of gravel blown out—the cone and remainder of the gravel remained.
139	2 0	2	2 oz. Govt.	Iron cone 3¾ in. then broken granite stone (about ½ in. cube) to top.	All blown out.

Experiments on Tamping with different kinds of Iron Plugs or Cones—continued.

No. of experiments.	Holes. Depth. ft. in.	Holes. Diameter. in.	Charge and description of powder.	Description of tamping.	Results and Observations.
140	2 0	2	2 oz. Gov^t.	Iron cone 2¼ inches, then broken stone (about ½ inch cube) to top.	All blown out.
141	2 0	2	1 oz. Contr^a.	Iron cone 6 inches, then broken limestone to top.	All blown out.
142	3 0	2	2 oz. Contr^a.	4 inches broken limestone, then iron cone, with broken stone to top.	All blown out.
143	2 0	2	1 oz. Gov^t.	5 inches clay, then iron cone 6 inches long, 1¼ inches diameter at base, then sand to top.	All blown out.
144	2 0	2	2 oz. Contr^a.	8 inches clay, then iron cone 5½ inches, then broken granite to top.	All blown out except about 2 inches of clay at bottom.
145	3 0	2	2 oz. Contr^a.	14 in. sand, then 4 in. clay, then iron cone 4 inches, with broken stone (about 18 inches) to top.	The cone, broken stone, and some sand blown out, as well as a small portion of the clay.
146	3 0	2	2 oz. Contr^a.	2 feet fine sand, then iron cone wedged with 3 arrows ⅜ inch thick, then sand to top.	All the sand blown out from *below* as well as above the cone, except about 2 inches, which was partly vitrified; the cone and arrows remained, and more firmly wedged.
147	3 0	2	2 oz. Contr^a.	Precisely as the preceding, but with broken stone instead of sand.	The stone above the cone blown out, that below remained; the cone more firmly wedged.

Experiments on Tamping with different kinds of Iron Plugs or Cones—continued.

No. of experiments.	Holes.		Charge and description of powder.	Description of tamping.	Results and Observations.
	Depth.	Diameter.			
No.	ft. in.	in.			
148	3 0	2	4 oz. Gov^t.	2 feet 2 inches broken limestone, with fine sand intermixed, then cone and 3 arrows, then stone and sand to top.	Stone and sand over the cone blown out; the cone more firmly wedged. It was conceived that the intermixture of fine sand with the broken stone might prevent the explosion escaping round the side of the cone; but it did not.
149	2 0	2	2 oz. Gov^t.	12 inches fine sand, then iron cone 4 inches, with 3 arrows.	The sand all blown out, except about an inch; cone and arrows firmly fixed.
150	2 0	2	2 oz. Contr.	14 inches sand, middling quality, then 4 in. clay, then iron cone, with 3 arrows, and sand to top.	Upper sand blown out—cone wedged very firmly.
151	2 0	2	4 oz. Gov^t.	15 inches sand, then 4 inches clay, then cone, with 3 arrows, and sand to top. N.B.—Only 3 inches for cone and arrows.	Upper sand blown out—cone remained fixed. The clay (4 inches) under the cone was found to be detached from the side of the rock for about half its circumference, the middle of which part was where the fuse had passed.
152	2 0	2	2 oz. Gov^t.	14 in. fine gravel, then cone 4 in. with 4 arrows only ¼ in. thick.	All blown out—the arrows in this experiment were too thin to wedge firmly.
153	2 0	2	2 oz. Gov^t.	12 inches of small broken stone and its dust, then cone 4 inches, and 3 arrows.	All blown out—this instance of failure could not be accounted for—must have been occasioned by some accidental circumstance—*vide* next experiment.
154	1 3	2	2 oz. Contr.	5 inches broken stone, then cone 5¼ inches, and 3 arrows.	Cone raised a little and firmly wedged—the explosion passed round the cone, being observed to escape at the top.

Experiments on Tamping with different kinds of Iron Plugs or Cones—continued.

No. of experiments.	Holes. Depth.	Holes. Diameter.	Charge and description of powder.	Description of tamping.	Results and Observations.
	ft. in.	in.			
155	2 0	2	2 oz. Gov^t.	6 inches clay, then cone 6 inches, and 3 arrows.	All blown out—one of the arrows found greatly bent.
156	2 0	2	2 oz. Contr^s.	6 inches clay, then cone 5½ inches, and 3 arrows.	Cone not stirred—nearly all the clay was found to be blown out from *under* the cone.
157	2 0	2	2 oz. Gov^t.	8 inches clay, then cone 6 inches, with 3 arrows.	Cone and arrows thoroughly wedged—rock uninjured.
158	2 0	2	2 oz. Gov^t.	10 inches clay, then cone 6 inches, and three arrows.	Cone and arrows thoroughly wedged—rock well cracked.
159	1 2¼	2	2 oz. Contr^s.	4 inches clay, then cone, with 3 arrows.	Cone and arrows thoroughly wedged.
160	3 0	2	1 lb. Contr^s.	12 inches clay, then cone 4 inches, with 3 arrows.	Rock thoroughly split, and large mass lifted—the clay remained, but opened by splitting of rock—the cone found inside the hole, the arrows thrown out.
161	2 0	3	4 oz. Contr^s.	6 inches clay, then cone 6 inches, and 4 arrows.	All blown out—broken parts of the arrows found flattened, and side of the hole marked with the arrows—the *number* of arrows evidently insufficient for the 3-inch hole.
162	2 0	3	4 oz. Contr^s.	8 inches clay, then cone, with 4 arrows.	All blown out—arrows found much bent and flattened.
163	2 0	3	4 oz. Contr^s.	13 inches clay, then cone, with 5 arrows.	The cone well wedged—about 2 inches of clay removed.

The conclusions which I think may be reasonably drawn from the foregoing experiments and observations on tamping, are,

1. That clay dried to a certain extent is, all things considered, the best material that can be used for tamping.

2. That broken brick, tempered with a little moisture during the operation, is the next best material.

3. That some kinds of rotten stone are as good as either, but that it is not so easy to be sure of always having the proper kind, and the use of it is very likely to lead to an occasional substitution or mixture of stone of other quality, such as is decidedly objectionable.

4. That sand, or any other matter poured in loose, is entirely inefficient.

5. That the stone dust and chippings of the excavation itself (excepting the rotten kind above mentioned) afford less resistance than clay, and being always more or less attended with risk of accidents by untimely explosion, should never be employed.

6. That of the mechanical contrivances by means of plugs or wedges, the most effective of those referred to are, the cone with arrows, and the barrel-shaped plug; both of which, particularly the former, give a great increase of resistance; but that all such contrivances, leading to increased expense, requiring extra arrangements, and some attention to a proper application, such as cannot always be depended upon, are none of them applicable to ordinary purposes, but might be very useful under circumstances where every blast is under great difficulties, or attended with much expense; for instance, under water, or in carrying on shafts, galleries, &c., through very hard rocks; in such cases the additional cost and labour of employing these means would perhaps be well repaid by the improved effect of each explosion.

In the case of blasting in shafts and confined driftways, where the openings made by the explosion *must* necessarily be across the line of the hole, one great difficulty, namely, that of disengaging the cone or plug after the firing, does not occur.

In very confined situations, like the bottom of small shafts, where the greatest possible effect is required from the shallowest holes; the effects from deep holes being counteracted by the infinite resistance of the contiguous masses, these means of increasing the resistance of the tamping might be very useful.

The comparative strength of different kinds of tamping will vary in a small degree in different places; that which is most commonly used, will, in each locality, give, relatively, no doubt, better effects; thus where broken brick or broken stone are commonly employed, they will perhaps on trial be found to give somewhat better results, as compared with clay, than are noted in these Tables, which were drawn up where the clay was in habitual use, and the others were only tried for experiment; but it is apprehended that, although they may vary in degree, in no case would the difference be so great as to alter the relative *order* in which they stand.

TUNNELLING.

By far the most difficult, expensive, and dilatory blasting operations, are those connected with sinking shafts and driving galleries in rock.

The disadvantages under which such work proceeds, arise from—

1. Want of space in which to work to most effect.
2. Want of light.
3. Want of pure air.
4. Penetration of water into the works, sometimes in large quantities.
5. The necessity for securing the parts excavated, on every side, from loose fragments, or other portions that might give way.
6. The inconvenient communication for men and implements, for removing the material, &c., to and from the work during the operation.

These peculiarities are such as to justify and even to require the most perfect arrangements and contrivances to be made use of; some that might be considered for ordinary blasting, as leading to unnecessary refinement and cost.

The easiest tunnels or galleries to open, are those that can be worked from the ends, without requiring any shaft; and more particularly those worked on an ascending inclination, by which any water met with has a natural drainage away.

Sometimes it may be so well circumstanced, as to rise from both ends, and have a summit in the interior.

The disadvantage of blasting in a re-entering angle, surrounded on all sides by great resistance, has been adverted to: this would be the case nearly through the whole progress of a tunnel, if the face were all to be taken out together, and therefore the following, it is conceived, would be a better order of proceeding, and has partially been acted upon in some instances.

Commence by a gallery or driftway A (fig. 17), at the apex of the roof,* as small as can conveniently be worked in; viz., about 6 feet high by 3 feet wide. This will be carried on necessarily in the most disadvantageous way for the effect of the powder, as previously explained.

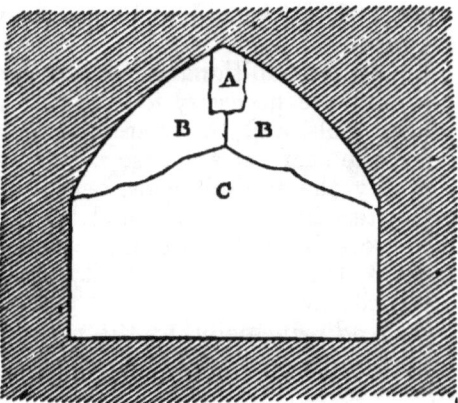

Fig. 17.

The next portion to work out will be B B; and in that some advantage will be obtained from the first opening A, to apply the powder to more effect.

The great mass C will be then removed almost as favourably as in an open quarry, having a face towards that part of the same portion C, that has been previously removed, to which the lines of least resistance can be directed.

Thus A will throughout be kept well in advance of B, and B in advance of C.

It is probable that each, without checking the progress of the part preceding, may be made to keep pace with it; and if so, the whole tunnel may be worked out in nearly the same time as would be required for a driftway.

The advantage of beginning at the roof is,—

1. The scaffolding and ladders for working from, become unnecessary.

2. The roof or arch is easily got at, and made sound and secure in the first instance.

3. The drainage is facilitated as the work proceeds, as will be explained.

In tunnels of very great length, the operation would be far too long by working them only from one or even from both

---

* The driftway is more usually carried along the *floor* of the gallery, to give a supposed advantage for drainage; but, as I think, erroneously.

ends: it becomes necessary then to sink shafts to intermediate points, from the bottom of which the tunnel is carried on in each direction until the whole shall be connected.

This of course will add very greatly to the difficulty and expense:

1. By the additional labour of excavating the shafts.
2. By the necessity for passing all the workmen, implements, tools, and the materials from the excavation, up or down the shaft, as the case may be.
3. By the limited space below, particularly as subject to the interruptions occasioned by the explosions, from which it is more difficult to obtain refuge.
4. By the probable necessity for increased artificial means of ventilation.
5. And principally, by the machinery and labour requisite to keep each distinct shaft, and the galleries connected with it, free from water.

The expenses will be increased in proportion to the number of shafts; and the number of shafts will be regulated by the time to which the execution of the tunnel is limited.

Thus, if the nature of the rock and other circumstances are such, that the engineer cannot be sure of penetrating faster, at each head of driving the gallery, than 3 yards per week, if the tunnel is to be 2 miles long, and the period for its being opened three years from the commencement; on these data 468 yards would be the amount that would be opened during three years in each head of working, requiring consequently between 7 and 8 heads for the whole, or three shafts, and the two ends. This, however, would be to suppose that the work commences at once from the *bottom* of each shaft: the time necessary for sinking the shafts, however, must be considered, which will consequently increase the necessary number of them.

A good record of the actual result of working tunnels, with a minute detail of all particulars, is very much wanted as a guide for future operations.

Many circumstances will have influence upon the dimensions to be given to tunnels; if for a railway,—the width of the locomotives and carriages, the height of the former, &c.

Certain dimensions may be deemed fixed: for a railway, for instance, the height of the vertical side walls suppose 10 feet; the space between the two tracks, and between each of them and the walls, suppose 16 feet in all; besides the width of the track of this space, it might be better to diminish that in the middle, and to increase those on the sides, suppose 4 feet for the first, and 6 feet for each of the others, because—

1. The side spaces would be more useful for gas or other pipes, or for lamps, or for drains, or for tools or materials during repairs.

2. It is the readiest and most natural position for any person to seek for refuge, should an engine or carriage pass while he may be in the tunnel.

The roof should be as flat as the rock will admit to be secure, provided there be sufficient height for the engine up to $a\,b$, (fig. 18,) over each trackway.

Some rocks will bear being cut to an almost straight horizontal roof; this will be particularly the case where the strata are vertical, or nearly so:

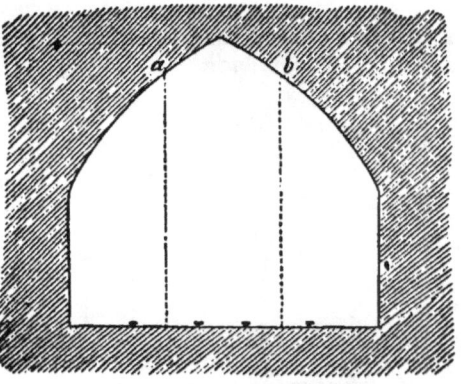

Fig. 18.

others require to be high and pointed; when even that form shall be insufficient the very troublesome and costly expedient of lining with brick or masonry must be resorted to

It has been sometimes proposed to construct a double gallery for a tunnel; one for the traffic in one direction, and the other for that in the opposite: the assumed advantages are—

1. That a small roof may be more easily secured than a larger one.

2. That one line can be opened first, and the second worked out by degrees, and at leisure, without interrupting the traffic in the first.

3. That the cross traffic cannot at any time interfere, and that persons who may be accidentally in a gallery while an engine or carriage is about to pass, will find secure refuge in the other, or in the connecting passages.

4. That the subsequent repairs, &c., will be quite safe from any interference with the traffic, which would then be confined to the line not under repair.

In chalk, a material that is easily worked and shaped, and in rocks like it, and in which a large roof may be of doubtful stability, while a small one might be considered secure, this system would be most applicable; but I should think it objectionable in hard rocks requiring to be blasted.

A single tunnel of 28 feet wide will afford as much useful accommodation as two of 18 feet wide each.

Suppose the width of trackway to be 6 feet; then, in either case, there would be a space of 6 feet between each trackway and the wall: but as it is proposed above to allow but 4 feet between the two trackways, it may be considered a more fair comparison to calculate each of the double galleries as only 16 feet wide.

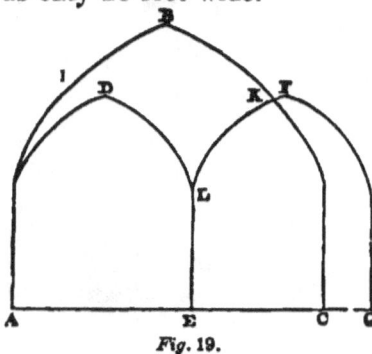

Fig. 19.

The comparison of the relative labour of opening either may be exhibited thus, bringing the two double galleries (which would be several feet asunder in practice) close together for the consideration of the question (fig. 19).

The excavation will in quantity be probably about the same, allowing for the connecting passages between the two smaller ones; but the single will be much easier to work out, because the space to work upon will be larger.

In the trimming work, which, after the first opening or driftway, is the most troublesome, the advantage is greatly in favour of the large gallery; as in the one case there is the amount of trimming the lines F, L, E, and D, L, E, besides the connecting passages, to compare with I, B, K, which they much exceed.

The single gallery will also require but one driftway; the double, two,—a point of great importance as regards labour and expense.

The single also gives additional space and height at the roof, where it is very useful as a receptacle for the vapours of steam and smoke.

It may be mentioned here, that the time and expense consumed in working out a shaft or gallery will be by no means increased in the direct proportion of its size.

One of reduced dimensions would take very nearly the same time in excavating as a larger (unless the difference be excessive), and the expense per cube yard will decrease

QUARRYING AND BLASTING ROCKS. 65

with the extent of opening,—the first or driftway of a few feet in either case being by far the most costly.*

Where a shaft or gallery is full large, there will be less occasion to be very particular in trimming the sides, by which much time and expense will be saved.

With regard to the floor of a gallery, no particular nicety is required, except to take care to excavate enough: if it should be more, it is of no consequence, as any hollows can be filled up without inconvenience, whereas in sides or roof any excess of rock removed may be troublesome to remedy.

The only assumed advantage of the double gallery worthy of mention, to compensate for the above-stated points of inferiority, is that of the capability of opening them in succession, the second, during the operation, not interfering with the first; but though it has an advantage in that respect, still it is apprehended that the difficulty of doubling the opening of a single gallery, one trackway being previously made good, would not be great.

Should it for any reason be judged inexpedient to open a double line at once, the mode will be to open the driftway 1 (fig. 20), and then one of the sides 2, and then 3 on the same side, leaving the other portion 4 for subsequent completion.

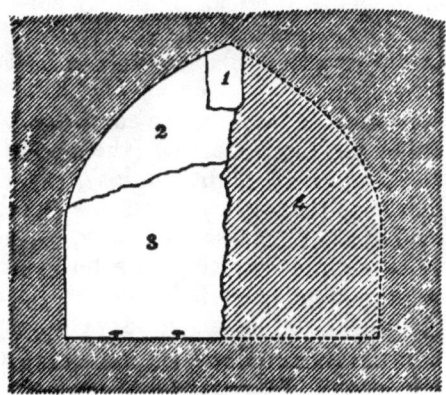

Fig. 20.

The half tunnel 1, 2, 3, will be more easy to open than one of the double galleries, on account of the *less extent of nice trimming*, &c., and the other half 4 will be far easier than the second small gallery.

The only inconvenience of this proposed mode will be, the danger of incommoding the first line of railway by the materials from the subsequent excavation; but it is apprehended that with moderate precautions this need not be the case.

The work will be gradually performed, the blasts not large;

---

* An example of the relative expense of this, and of shafts, &c., will be found in the cost of the tunnel of Drenodrohur, page 73.

they may be always fired immediately after the passing of a train, and the line cleared of any little rubbish before the next shall arrive, and a proper system of signals established, by which every train at its entry may be assured that the line is clear.

In the single tunnel, the lower end of the shafts will be in the middle, and form part of it; whereas in the double, however arranged, the shafts cannot be so conveniently placed with reference to both galleries.

SHAFTS.

In sinking shafts, the work is much more disadvantageous than even in the galleries, on account of—

1. The more limited space.
2. The more constrained position of the workmen.
3. The danger of anything, even small articles, falling down upon them.
4. The difficulty of applying any holes for blasting, but such as are vertical, or nearly so.
5. The immediate vicinity of the receptacle for the drainage water, from which it will be almost impossible to keep clear.

It will be found more advantageous probably to sink shafts rather long and narrow (the length crossing the direction of the tunnel at right angles), than circular or square; for instance, 16 feet by 9 feet may be better dimensions to give a shaft than 12 feet each way, provided always that the width is *ample* for working the buckets up and down. By this arrangement it will be easier to apply transverse bearers across the top for machinery; the space for the workmen will be more convenient; the pipes for drawing off the water, for ventilation, &c., will be more out of the way at one end, while the buckets for drawing men and materials, &c., up and down, may be at the other.

The upper surface edge of each shaft must be thoroughly lined, and secured from the possibility of anything falling down; the buckets and windlass of perfect description and arrangement to prevent such accidents, or the striking against each other in meeting, &c.

It is in blasting at the bottom of shafts that the shields described in another part might, it is conceived, be more usefully applied than in any other situation. The holes will be vertical, the blasts small, the shield always close at hand,

QUARRYING AND BLASTING ROCKS. 67

and the great labour and loss of time saved of removing the workmen to the top of the shaft at each blast.

When the shaft is sunk to the depth of the *driftway* at the *roof* of the galleries, those driftways may be at once commenced to save time; and the remainder of the shaft sunk while the gallery work is going on.

EFFECTS OF STRATA.

In stratified rocks, the direction of the strata in tunnelling will be of much importance.

The most favourable would be vertical, and at right angles with the direction of the line of tunnel, because when the drift A (fig. 21) is carried to sufficient extent, the holes for the subsequent blasts can be bored down the joints, and the explosion made to act in the most favourable manner.

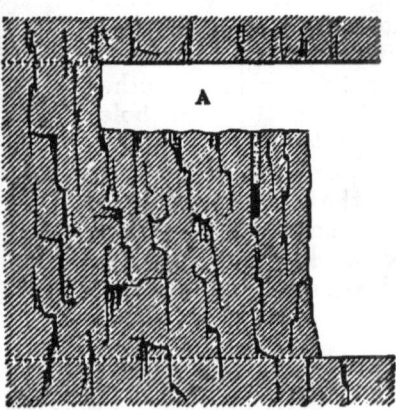

Fig. 21.—(Longitudinal Section.)

The same principle may be adopted where the strata may be inclined upwards from the miners, as at B (fig. 22), but it will be

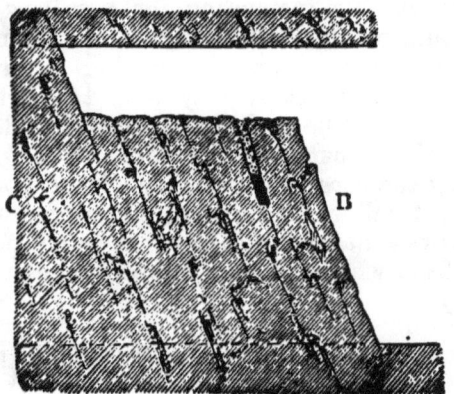

Fig. 22.—(Longitudinal Section.)

more difficult to work with advantage from the other end of the gallery c; as, to follow the same principle, the rock will be always overhanging the workmen.

If the strata be in horizontal beds, it will be worked upon by horizontal boring after the roof is entirely cleared out.

A most unfavourable direction is when the work is proceeding in the same line with vertical strata, which will consequently always present their edges in front.

In that case an opening should be carried down from top to bottom, either at one side at D (fig. 23), or in any other part, and then the holes bored down the strata as at A, or horizontally as at B; the whole roof, however, and opening D, will be worked to much disadvantage.

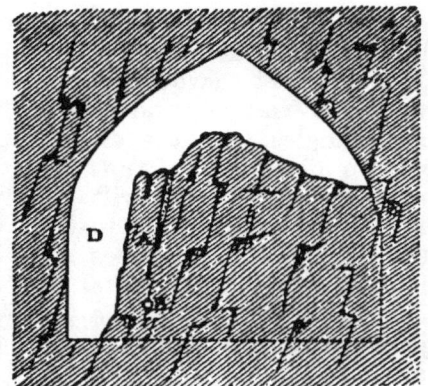

Fig. 23.

DRAINAGE.

An adequate drain will be necessary through the whole length of any tunnel.

A very favourable inclination for a gallery, as respects drainage, would be about 16 feet in a mile (1 in 330).*

The drain will be larger or smaller according to the quantity of water which it may have to discharge, but at 1 in 330 the fall will be sufficiently considerable to render a very large drain probably unnecessary.

In working the *ascending* galleries, either from the end or from a shaft, it will be quite easy to proceed so that there shall be always a natural drainage, at every period, to the receptacles from whence it will run off, or be pumped out, as the case may be.

---

* Whether for ordinary road or railway, the incline in a tunnel should be easy; as it would be peculiarly inconvenient in that particular part of a line to incur the risk of a stoppage in ascending, or of an accident in descending.

But in each *descending* gallery some little arrangement must be made to assist it.

When the descending meets the ascending gallery, the whole of the drainage will be carried off by the latter; till then, the following arrangement will be better than carrying pipes to the end of the gallery, as must be done if the inclination be steep.

Suppose the descending gallery to be 440 yards in length before it is to meet the other; this, at 16 feet to a mile, will give a perpendicular difference of level of 4 feet: as the gallery will be from 24 to 30 feet high, it is manifest that the far greater portion may have a regular natural drainage out, the same as an ascending gallery; but by sinking a small well-hole at the bottom of the shaft, of 5 or 6 feet deep, from whence the pump may draw the water out, a drain may be cut to it from the very end of such descending gallery, as it progresses, that will carry off the water naturally from its very floor.

That drain will form part of the regular longitudinal drain that must be made at all events, and the only additional labour will be in excavating the well in each shaft, which also might at any rate be desirable.

When the tunnel or gallery is to be horizontal, as it must be for a canal, the drainage will be as described for the *descending* gallery.

In deep sinkings, drainage is sometimes effected, or rather more usually *assisted*, by means of artesian wells or bore-holes sunk from lower levels of the surface ground, down to the lines of strata that shall *descend* from the works. Where the joints shall not be too close, and the distance not too great, the water will be drawn from the works by this self-acting drainage down to the level of the top of the artesian bore-hole; or, *practically*, to nearly that level.

VENTILATION.

The difficulty of working galleries to any extent under ground is occasionally very great, for want of ventilation, or from the presence of foul air.

The remedies are, either to force fresh air to the end of the work, or to draw the foul air off from thence, when the fresh air will rush into the vacuum: the latter is esteemed to be the more easily effected.

In either case there must be an air-tight tube from the

fresh air to the part where the workmen are engaged: and the great difficulty must be to establish such a tube perfectly firm and perfectly air-tight.

In operations on a small scale, such as military miners are engaged in, a small and light tubing is used, and the air forced in by a pair of smith's bellows, or others of about equal power; but to work a gallery of some hundred yards long from the bottom of deep shafts will require tubing more substantial and of larger dimensions, with a continued force applied for exhausting the foul air; or fresh air has frequently been introduced by a fan-wheel.

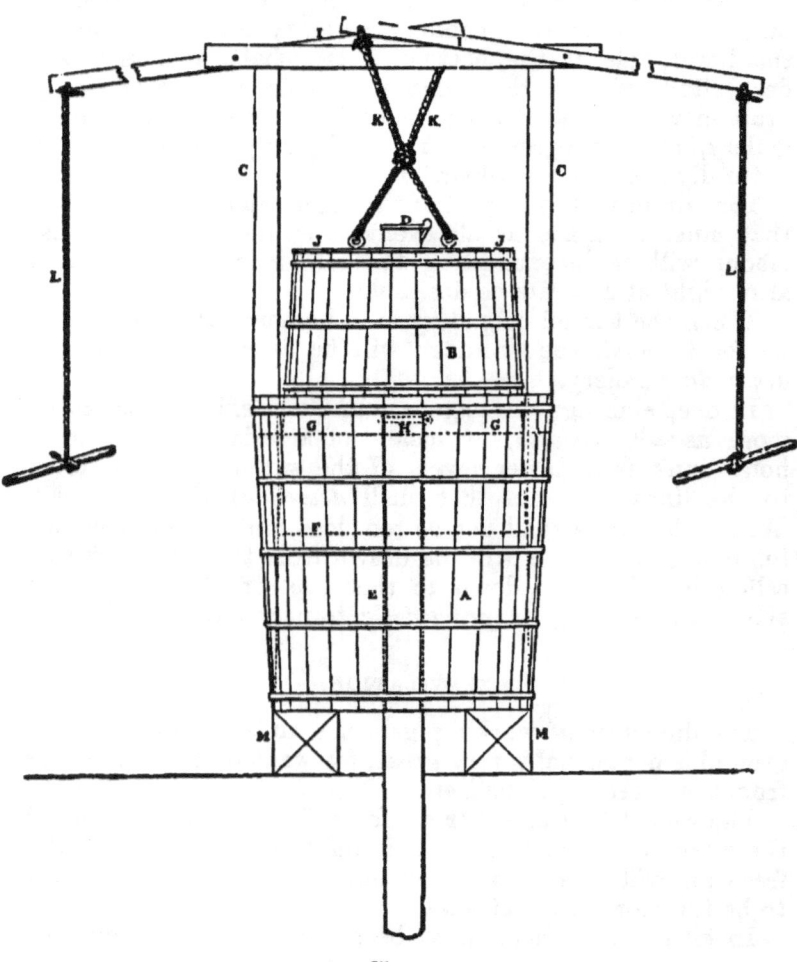

Fig. 24.

In a recent number of the Annales des Ponts et Chausées of France, there is an account of a simple contrivance which enabled the working of a shaft of 5 feet diameter, and 220 feet deep, to be continued after it had been interrupted by the constant collection of carbonic acid gas; all the ordinary measures by bellows, &c., having proved quite ineffectual.

A large tub A (fig. 24) was firmly placed on balks on a level with the top of the shaft, and filled with water to the level G, G.

An air-tight pipe from the bottom of the shaft was brought through the tub A, and had its upper edge a very few inches above the water; it had a valve H on the top.

A smaller tub B, reversed, was suspended within A, by the cords K K, which were made fast to the ends of the levers I I.

B had a very short pipe at top, with a valve D.

The tub B being allowed to descend by its own weight, the air within it was expelled through the opening D; when again raised by pulling the handles attached to the ropes L L, the air was drawn up through the opening H, from the end of the descending tube, and by continuing this reciprocating action, a circulation was created at the very bottom of the shaft.

No dimensions are given; but it is conjectured that the lower tub A would be about 4 feet by 3 feet 6 inches, and the upper one B, about 3 feet by 3 feet.

It was found capable of drawing off 4 cubic yards of air per minute.

No additional men were employed to work it; those at the top of the shaft, who got out the materials from the excavation, were only required to work this pump for about 5 minutes every hour, to keep the air perfectly good at the bottom.

It is in fact an air-exhausting pump of large power and simple construction: if found useful, and required for regular service, it might no doubt be improved, and made more compact, portable, and easy of application.

Instead of using a pump or manual labour for drawing the air from the ends of galleries, the upper ends of the tubes are sometimes made to communicate with a fire kept constantly burning, to which the tubes furnish exclusively the requisite air.

Improved ventilation is said to be obtained by double galleries or driftways, with occasional air-holes connecting them.

The advantage in this respect can hardly be equal to the cost of carrying on the double opening, if not necessary for other objects.

COST.

No kind of work can be more variable in its cost than tunnelling, or driving galleries: so much will depend upon the quality of the rock, the depth at which the gallery is under ground, the facility of communication by more or less of confined galleries, shafts, &c., to and from the actual work,—the amount of necessary ventilation, and the quantity of water to be drawn off,—that the expense may vary from 2s. to £2, and much more, per cube yard.

Where the rate is very heavy, it will probably be good economy to adopt the most refined improvements that will forward that portion of the operation which peculiarly leads to the extra expense.

The following is the result of the cost of working out the road tunnel of Drenodrohur, between Kenmare and Glengariff, during 1836 and 1837, in the county of Kerry, under many favourable circumstances.

The piercing of this tunnel, which was 582 feet in length, through the summit ridge of the mountain, saved an ascent, of which the perpendicular height was from 60 to 80 feet, which, reduced from a constant rise of 1000 feet on each side, was of much value.

The rock was stratified, varying in character from a granular to a compact silicious, and to a common clay slate, intersected by veins of quartz. It was all hard, some of it exceedingly so.

The strata in one direction were nearly vertical to the horizon, excepting occasional veins of a few feet in thickness, which were more inclined; and in the other direction, nearly perpendicular to the line of the tunnel.

The rise in the tunnel was from both ends, and at the rate of 1 in 100, forming a summit in the middle, which was consequently nearly 3 feet higher than the entrance.

The passage, as commenced, was meant to be 18 feet wide and 18 feet high; but subsequently, to save expense, the width of the roof or chord of the arch was reduced to from 15 to 17 feet, and the part below the arch to 12 feet.

The arch stood perfectly without any support, although cut *very* flat, and with a rise or versed sine of only 2 feet.

The cost of the different portions was as follows:—

The roof or arch was first cut out, and contained sectional areas at different parts of about 96 and 90 feet, comprising an excavation of about 2012 cube yards in the aggregate, and cost £6555 16s. 8d., being at the rate of about 6s. 6d. per cube yard.

The body of the gallery under the arch contained sectional areas of about 204 and 144 feet, comprising an excavation of about 3493 cube yards, and cost £599 1s. 11d., being at the rate of about 3s. 5¼d. per cube yard.

A small shaft, 6 feet by 4 feet, was opened to the depth of 33 feet, containing about 29¼ cube yards, and cost £32 2s. 10d., being at the rate of £1 1s. 10¾d. per cube yard.*

The average cost of tunnel alone per cube yard 4s. 6½d.
" " " " and shaft . . . 4s. 8d.

The material from the excavation was deposited in hollows near the entrance at each end.

The total cost of tunnel and shaft £1287 1s. 5d., exclusive of some miners' cottages built, and a few such contingencies.

HEADS OF EXPENSE.

		£	s.	d.
Labour . .	Miners and labourers	620	9	8
" . .	Smiths (including coals	73	3	3
Iron . . .	13,109 lbs. at 1s. 7d. per lb. . . .	92	16	6
Steel . . .	321 lbs. at 1s. per lb. . . .	16	1	0
Gunpowder .	7946 lbs. at 1s. per lb.	397	6	0
Safety fuse .	1745 coils, of 8 yards each, at 1s.	87	5	0
		£1287	1	5

At the railway tunnel at Liverpool, the excavation of the rock is stated to have cost 4s. per cube yard; the material was a sandstone, and, as it required to be lined nearly throughout with brick-wall and arch, was not probably very hard.

It would form a most useful guide to engineers, if record were kept, and made public, of all the particulars of the actual experience in tunnelling and driving galleries, and sinking shafts: it might be kept without difficulty, and be made to afford very useful checks on the works during the operation.

J. F. B.

* This shaft proved eventually of little or no use, and had better have been omitted.

APPENDIX.

In the following Memorandum, Major-Gen. Sir John Burgoyne describes the mode in use at Marseilles for enlarging the inner end of the hole by means of the action of acid upon the calcareous rocks of that district.

The process for blasting calcareous rocks at Marseilles is extremely simple; it is merely to form a chamber or magazine at the bottom of the drift by means of dilute acid, so as to admit of a larger charge of powder being employed for the blast than is available in the ordinary mode of proceeding, the rock being of a calcareous nature. The apparatus was somewhat rude, but it appeared to answer its purpose very effectually. The rock was pierced in the usual way, by a $2\frac{1}{4}$ inch jumper, to the depth of 5 to 7 feet, generally in a slanting direction, according to the form and mass of the rock to be detached. They then introduced a copper pipe, the size of the bore, in the form sketched on the margin, and pressed the end A, which is open, down to the bottom of the hole, the orifice round the outside of the pipe at B being closed up tight with clay, so that no air could escape; and the bent neck of the pipe c, which is open, hanging downwards, with reference to the slope of the bore. Through the copper pipe at D was introduced a small leaden pipe, e, of about $\frac{1}{2}$ inch diameter, formed with a funnel, f, at the top, and this passed down through the copper pipe to within about an inch of the bottom, the upper orifice of the copper tube round the leaden one at g being filled up with a packing of hemp.

Matters being thus adjusted, dilute nitric acid was poured through the funnel and leaden pipe, and, on arriving at the bottom it produced effervescence, and at the end of a few minutes the frothy substance of the dissolved rock began to run out through the orifice of the bent tube at c. So long as they poured in the dilute acid at f this action continued, and, whenever they considered from the quantity of substance delivered through the pipe that the internal chamber had become of sufficient magnitude to admit the quantity of gunpowder required, the pipes were withdrawn; the hole very soon became dry enough, from the very action of the acid upon the limestone, to permit the chamber being charged without damage to the powder, and the blast was given in the usual way.

The time occupied would of course vary according to the size of chamber required. The foreman of the works informed me that he had succeeded in forming a chamber capable of holding 25 kilos. of powder (about 55 lbs. English) in the space of 4 hours; he showed an unwillingness to give me the proportions of acid that he used, or the nature of it, and I did not therefore press my questions, being satisfied by the vapour that it was nitric acid, and the strength of the mixture must in all cases vary according to the greater or less proportion of calcareous matter existent in the rock to be acted upon. This may at all times be determined in a few minutes by experiment upon pieces of the stone, with acids of different degrees of strength.

They were blasting very large masses from the foot of the hill, on which the Fort of Notre Dame de la Garde stands, on the south side of the town, for the purpose, in the first place, of obtaining sites for building, and at the same time making use of the *débris* for forming the new port which is in progress, and the foreman assured me that this new process had operated a very large saving to the MM. Lerm, Frères, who are the parties that have brought it into action.

In March, 1845, a patent was granted to William Joseph Conrad Marif, Baron de Lièbhaber, of Paris, for " Improvements in blasting rocks and other mineral substances for mining and other purposes, and in apparatus to be used in such works." The inventor's mode of enlarging the inner end of

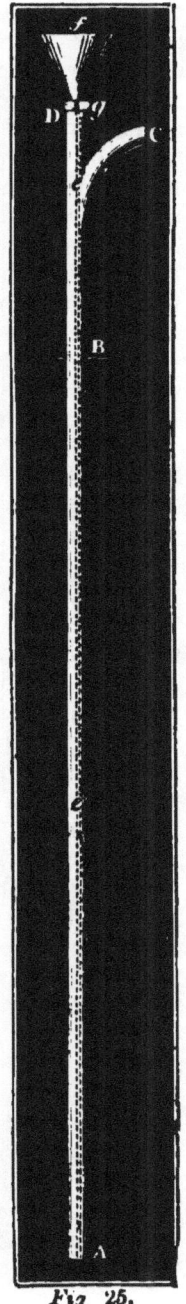

Fig 25.

the hole was described to consist in dissolving a portion of the stone by means of muriatic or other acid, diluted with about three times its weight of water. A tube is inserted in the hole and externally sealed round at the lower end with some composition which shall prevent the froth or vapours from the acid passing between the outside of the tube and the inside of the hole bored in the rock. Within this tube there is a smaller tube, through which the acid passed into the hole. These tubes are bent over at the top, and terminate in a vessel containing the acid, and which vessel receives the froth that passes up between the pipes. The inner tube is bent upward at the lower end so as to prevent the froth passing through the stone. When the hole is sufficiently enlarged, the contents of the hole are removed by a siphon or pump before drying and charging with the powder.

In addition to the important facts given by Gen. Burgoyne, the following regarding some operations in blasting in the Jumna and at Delhi, conducted by Lieut. Tremenheere, may be quoted as useful and interesting. For the purpose of blasting in order to improve the navigation of the Jumna, the jumpers used were 6 feet long, and $2\frac{1}{4}$ inches in diameter; the blasts 5 feet deep, and 4 feet from each other. The rate of boring varied from $2\frac{1}{2}$ to 5 feet per day's work for 2 men. A double-headed jumper was used, to render the hole completely circular for the reception of the canister, about $2\frac{1}{2}$ feet in length, and 2 inches in diameter, and filled two-thirds with powder and the rest with sand. The small tube reaching to the surface of the water contained quick match with a piece of slow match at the extremity. The canister, well greased, was placed in the hole without any additional tamping.

At Delhi, the blasting was in dry rock, and, economy of powder being of more importance than economy of time, tamping was resorted to. For this a stiff red clay, slightly moistened, was employed, and the tamping bar was of wood, the priming wire of copper. Any dampness which might exist in the bore was obviated by a tube of coarse paper, greased on the outside. Fine mealed powder was used as priming, and a piece of port-fire for ignition. If the firing did not succeed, a fresh priming hole was bored in the tamping, or the mine abandoned. In large irregular masses of rock, the depth of the bore, or the intervals between the blasts, will generally represent the line of least resistance; and the following results were obtained in the rock at Delhi, which is hard quartz. The line of least resistance not exceeding

1 foot, a charge of 2 oz. is sufficient; the line not exceeding 4 feet, and the rock not being highly crystalline, 3 oz. per foot will be sufficient. The charges will vary with the tenacity of the rock, but the following may be a general guide:— the line of least resistance being 1, 2, 3, 4, 5, 6 feet, the charge will be 4, 8, 14, 20, 26, 36 ounces. On comparing the charges used at Delhi, where stiff clay was used as tamping, with those in the Jumna, where sand was used, the following table is the result:—

Line of least resistance.	With clay tamping.	With sand.
2 feet	8	26·8 oz.
2½ ,,	10	33·5 ,,
3 ,,	12	40·2 ,,
4 ,,	20	53·6 ,,

The charges in the last column are to those in the second as 3 to 1, nearly; they are not, however, given as the least required, but are those actually used.*

In the neighbourhood of Plymouth, the limestone, with which that district abounds, is raised in solid masses, of from 3 to 10 tons' weight: it is used most extensively for building, and for lime manure. About 13 cubic feet weigh a ton. It is of a light blue or grey colour, generally free from metallic veins, but with some indications of manganese and ironstone, round pieces of the latter being found in clay beds, intermixed with the rock, and a vein of ironstone 4 inches thick at the surface of the rock, and dipping towards the south, has been opened.

In the blasting of this stone, if a deep hole is required, a 2-inch jumper or bit is used for about 4 feet, and a 1⅞ inch for the next 4 feet by one man; then two men are employed with a 1¾-inch to the depth of 14 feet, a 1⅝-inch to the depth of 21 feet. A constant supply of water is required during the boring. The hole being well dried, about one-third is filled with powder, say 15 lbs.; a needle is introduced as far as possible without driving it; the hole is tamped with dry clay to the top, and then covered with a little wet clay to prevent any of the loose particles falling in when the needle is withdrawn. A reed filled with powder and split at the top, to prevent its falling to the bottom of the hole, is inserted, and a stone laid upon it; the powder being ignited by a piece of touch-paper and a train, the reed flies to the bottom of

* Report of Papers, Inst. Civ. Engineers.

the hole, and ignites the main load. The rock is generally cracked and loosened to a considerable extent, if not thrown; in that case the needle is driven through the tamping, and such a fresh charge is run through the needle-hole as may be requisite. From 6 to 8 tons of rock are generally blasted with 1 cwt. of powder. In July last, however, it was reported that with one charge of 1 cwt. of powder, in a hole 18 feet deep, no less than 1000 tons of the limestone rock, destined for the Plymouth Breakwater, were detached from their bed. We cannot answer for the entire truth of this report; but it is well known that, since the Breakwater quarries were opened in 1811, a great augmentation has been effected in the produce of the blasting there carried on.

In blasting the white limestone of the Antrim coast, in the north of Ireland, it was found that one ounce of powder would rend 14·12 cubic feet of the stone, when in blocks; but the same quantity rent only 11·75 cubic feet of loose whinstone blocks. The specific gravity of the white limestone is nearly 2·76, of the whinstone or basalt about 3·20. The limestone is similar to the chalk of England, in the flints which it contains; but it is exceedingly indurated. The induration may be estimated from the fact that two men will bore one foot in depth in half-an-hour, with jumpers from 1¾ to 2 inches in diameter.

The following account of the method pursued in getting the valuable granites of Scotland is written by a gentleman of extensive practical knowledge of the subject.

The granite quarries of Aberdeen and Peterhead, in the north of Scotland, are comparatively inexhaustible, and the granite which they yield of excellent quality. The manner of extracting the rock in these quarries is by boring, and blasting it with gunpowder. In this process the quarry-men use the ordinary "jumpers," having a piece of steel welded to one end, and made with an edge similar to that of a stone-cutter's chisel, but much more obtuse. They are also made of different lengths—the long ones to follow the short, as the boring proceeds. Three men are employed in this department of the work—one sitting on the rock and turning the "jumper" round, while the other two strike alternately with hammers, each of which is about 12 lbs. weight. These bores are sometimes from 12 to 16, and often 18 feet deep, by about 2½ inches in diameter; and, when charged and fired, they generally loosen a very great quantity of rock, but seldom do much more than merely move it. "Bulling" is then had recourse to, and this consists of filling all the perpendicular cracks or fissures, caused by the former operation, with gunpowder, and firing it; by which means the rock is shifted several inches from its bed, and often thrown some yards forward. The next process is to cut up the rock into the required scantlings, and this is done in the following manner:—A number of holes are cut in a straight line, about 3 inches long, 2 deep, and 3 inches apart, each tapering towards the bottom, into which iron wedges are inserted, and struck with a heavy iron hammer, beginning at one end of the row and striking them from end to end until the block gives way.

Cutting with "Plug and Feather" is also had recourse to when the block intended to be split is very deep and supposed to be beyond the power of the ordinary wedges. In this case a row of circular holes, about $1\frac{1}{4}$ inch in diameter, 5 or 6 inches deep, and 6 inches apart, are bored with a "jumper;" into each of these are put two "feathers" at opposite sides. (The "feathers," when in position, are merely inverted wedges, having circular backs, so as to fit to the curvature of the holes, the opposite sides being plain, to receive the wedges.) Between these "feathers" long wedges are then introduced, and driven, as in the former case, until the block is split asunder.

In the Seventh Volume of the Professional Papers of the Corps of Royal Engineers,[*] Capt. Nelson, R.E., has recorded some interesting and useful memoranda on the quarrying of the Plymouth limestone; from which paper some facts may be well quoted. This limestone is described as fine-grained and hard, abounding in organic remains, and varying in colour from light to dark grey. The tools used in these quarries consist of—the "jumper," or "bit," from 8 to 30 feet in length, and from 1 to $2\frac{1}{2}$ inches in width;—the "pitching bar," from 6 to 8 feet long;—the "tamping bar," which is $\frac{7}{8}$ inch diameter for $1\frac{1}{4}$ inch bit;—the "needle," which is from 3 to 13 feet in length;—and the "ripping" or "crow bar," from 6 to 14 feet long, and from $2\frac{1}{2}$ to 3 inches square at the lower end. In addition to these the "feather and tearers" consist of a wedge about 6 inches long, 1 or $1\frac{1}{4}$ inch square at one end and $\frac{1}{4}$ inch square at the other end, and two thin wedges having one flat face and the other segmental; so that the central "feather" with two "tearers," one on each side, serve to fill pretty well a round hole in the

---

[*] Weale. 1845.

stone. The quarry-carts are short and very stoutly built, lined with strips $2\frac{1}{2} \times \frac{1}{2}$ inch flat iron; they carry from $1\frac{1}{2}$ to $1\frac{3}{4}$ tons—they are 4 feet long, 3 feet 6 inches wide at the hind end, and 3 feet 4 inches in front; the sides about 1 foot 6 inches high, and the front much higher; they cost about 15*l.* each, and, with repairs, will last nine years. The navigator-barrows, which are of the common pattern with wooden wheel, cost 10*s.* 6*d.* In the quarry, where they are used all day, and loaded and worked quietly, they are entirely of wood; but at the wharf, where the work is shorter and sharper, a body of plate iron is used on a wooden carriage. Iron carriages for the barrows would be more economical, but too heavy for the quarry-men. Iron wheels are considered too destructive to the planks. The barrows last from three to five months, according to the work, and without repairs. The planks, which are of yellow pine, and in the quarry generally about 25 feet by 14 inches by $2\frac{1}{2}$ inches, hooped and bolted, cost $4\frac{1}{2}d.$ per superficial foot. At the wharf the planks used are longer, broader, and thicker; they last about six months. The "jumpers" cost 5*d.* per lb. For the removal of heavy stones the "devil cart" is used; this consists of a pair of large and strong wheels with a stout and well-braced axle, to which a long central pole is fixed; the front and lower end of which is carried upon a small wheel or roller. To this cart the heavy stones are slung by double chains, one of which is attached to the hinder side of the axle, and the other to the pole as near to the axle as may be necessary, according to the dimensions of the stone carried. The quarry screw-jack is an implement of great power, resembling the ordinary tool of that name, but having a more extensive series of gearing within; the winch is on a pinion of four teeth working a wheel of fourteen; on this is another pinion of four teeth, working a wheel of eighteen, on which is a three-teethed pinion or triple stud engaged with the vertical ratchet.

Capt. Nelson's memoranda refer to three quarries in the neighbourhood of Plymouth, in the two first of which the limestone is chiefly raised in blocks of from $\frac{1}{4}$ to 1 cwt., either to be burned on the spot for lime, or shipped for different places in the neighbourhood for that purpose. The stone is likewise wrought for curb-stones and paving slabs, but cannot be got in masses large enough for wharf ashlar. In the third quarry the stone is raised in the largest blocks with heavy charges, and used for breakwaters and wharf ashlar. The largest block known to have been raised in this quarry weighed about 200 tons; and so variable are the effects of blasting, that, on one occasion, upwards of 1000 tons were detached by a charge of 100 lbs. of powder in a 30 feet hole. It is usually considered that, well stratified as the rock in this quarry is, where the mass to be raised is not jammed at the sides, and where the beds dip to the front of the quarry, a quarter of a pound of powder per ton will dislodge the block sufficiently.

In quarry No. 1, one man raises 5 tons per day, besides loading and unloading the cart, and removing the rubbish produced 50 yards. Wages (in 1839?) 15s. per week, on piece work. In jumping, 16 feet, in four 4-feet holes, is considered a good day's work. Powder used, about 4 lbs. for 16 tons; one-fourth of the hole filled with the charge.

In quarry No. 2, one man in one day (10 hours net time) will put down 20 feet with $1\frac{1}{2}$-inch bit, or 15 feet with 2-inch bit; depth of the charge from one-third to one-half that of the hole; half a pound of powder allowed per cubic yard of solid rock. A man will turn out and break up for rubble stone 4 tons, or 56 cubic feet per day; or 6 tons without breaking. When thus broken into pieces of from $\frac{1}{2}$ to $\frac{3}{4}$ cwt. each, and delivered at the kiln (close to the quarry), he will break it up for burning at one load of 16 tons, or 224 cubic feet per day.

In quarry No. 3, 20 feet of $1\frac{1}{4}$ to $1\frac{1}{2}$-inch bit is reckoned a good day's work, although one man puts down 30 feet, which is very unusual; in deep holes, one man is employed for the first 6 feet, two for the next 10; and, after 16 feet, three men, provided the hole is not more than 10° or 12° out of the perpendicular; if it is, even five or six men are put on after the first 15 feet. In such holes, it is usual to commence with a $2\frac{1}{2}$-inch bit, and end with about $1\frac{3}{4}$ inch. In a 30 feet hole, $2\frac{2}{3}$ inch in diameter, 25 lbs. of powder are used for the first charge. If this does not succeed, it will, at all events, probably disturb the beds beneath sufficiently to allow space for the repetition of the blast, perhaps three or four times over, increasing somewhat every time before the final charge is introduced; as the block, if not well detached on all sides first, will be destroyed by the premature application of the full charge.

SLATE QUARRYING.

In order to understand the peculiar methods adopted in

slate quarrying, it is necessary to refer to the distinguishing geological feature of the slate formation. Slate, besides being divided by fissures which form beds and joints to the masses, as is the case with all stratified materials, is divisible into laminæ or plates of any required thickness, the direction of which laminæ is commonly vertical, or nearly so, but always oblique to, and never coincident with, either the beds or joints. This peculiarity of structure affords the means of splitting the slate and adapting it so suitably to the extensive purposes for which it is employed in great Britain, and, at the same time, determines the most convenient manner of quarrying to be by detaching the masses of slate vertically from the face of a trench or gullet. The cutting of this gullet into the side of the slate mountain is, therefore, the first operation in the working of the quarry. As the trench proceeds, and the height of the surface above becomes greater than convenient (say about 40 feet), a second trench is commenced above the other, and similarly carried onward into the mountain until the height above reaches a similar quantity, when a third trench is commenced, and so on. In the great slate quarries of Bangor,* sixteen of these stages are in progress together; the lower ones being gradually widened, by the getting of the slates, as the upper ones are advanced. In the upper part of the quarry, the slates are removed with crow-bars; but the slates become harder as they are lower from the surface, and require the use of gunpowder to detach the main masses. The miners engaged in drilling the holes for the powder are suspended by ropes from the upper parts of the rock, and are liable to many and severe accidents. After the slates are detached by powder or otherwise, they consume considerable labour in splitting them with wedges and mallets into marketable sizes, and reducing them to the several sizes required for roofing and other purposes. The means of communication between the several stages or levels (each of which in the Bangor quarry

* These quarries are about six miles south-east of Bangor, and are worked in the mountain called in Welsh *Y Bron*, a name signifying a *breast* or *pap*, and commonly given to any hill the contour of which has a flowing outline, free from abruptness. This mountain, Y Bron, is on the side of a deep valley, through which the river Ogwen flows towards Lavan Sands, where it falls into the sea nearly opposite Beaumaris. An interesting paper on these quarries, to which we are indebted in preparing this notice, is contained in the Quarterly Papers on Engineering, vol. iii., part 5.—(*Weale*, 1845.)

is about 40 feet high, or nearly 640 feet from bottom to top) are provided by self-acting inclined planes, of great extent, laid from each stage to the contiguous stages. On these planes rails are laid to facilitate the motion of the trucks in which the slate is conveyed. At the head of each plane, a drum and break-wheel are fixed to regulate the velocity of the descent of the loaded trucks towards the level below. From the quarry at Bangor the slates are conveyed upon a railway to the shipping-port, a distance of about seven miles. On the several stages long ranges of sheds are erected for the men employed in cutting and shaping the slates. The slates for roofing are merely split to the required thickness with long iron wedges, and then trimmed with the cutting knife, by being placed on a fixed steel edge.

The dykes of greenstone, with which slate rocks are frequently traversed, are found to injure the slate, and destroy the cleavage of its structure for a considerable breadth. These dykes are usually removed by blasting, and form a very expensive impediment to the operations of the slate-miner.

At the Bangor quarries, about 1000 men are commonly employed, and the profits are understood to amount to about £80,000 per annum.

Some of the slate quarries of France are worked by subterranean galleries. In those of Rimogne the main gallery is about 400 feet long, with several lateral galleries, 200 feet in length, on each side of the main gallery. These galleries are about 60 feet high, and contain about 40 feet in height of good slate, the remainder consisting of quartz, or being injured by contact with volcanic substances.

At Angers, in France, are other slate quarries, worked in open cuttings, and which afford blocks, of which most of the houses of that town are built. In the German States there are several slate quarries: at Eisleben, in Saxony; at Ilmenau; Mansfield, in Thuringia; and at Pappenheim, in Franconia. Switzerland is said to possess no slate, except in the valley of Sernst, in the Canton of Glaris. Italy has only one quarry, that at Lavagna, in Genoa, the slate from which is used for lining the cisterns in which olive oil is kept.*

---

* (Papers on Engineering, vol. iii., part 5.) The application of slate here stated in the text is a good proof of the close texture and comparative impermeability of slate. M. Vialet states that he is able to double the natural hardness of this material by baking it in a brick

FIRING MINES BY VOLTAIC ELECTRICITY.

The removal of a large mass of the chalk cliffs near Dover for the works of the South-Eastern Railway Company, by a grand blasting operation effected by three mines, simultaneously fired by electrical communication, was a work of such surpassing magnitude, and so completely successful in its results, that a brief account of the arrangements made and effects produced may be suitably introduced in an Appendix to the valuable treatise of Sir J. F. Burgoyne. We have at the same time the greater pleasure and greater facility in doing this, because the operations have been so well described by the able officer under whose experience they were conducted,* and from whose description the following notice has been abridged.

The Round Down Cliff is situated 2 miles W. of Dover. Its summit is about 380 feet above high-water mark, and 70 feet above that of the Shakspeare Cliff, which is ¾ of a mile nearer to the town. It is composed of a compact chalk, principally without flints, forming the lower bed of the chalk formation. The dip inclines from W. to E. about 40 feet in a mile. The base jutted out to seaward considerably beyond the general line of cliff to the westward, under and along which (from the Abbott's Cliff Tunnel towards Folkestone) a distance of 2 miles of the South-Eastern Railway is passed on an open embankment, defended by a substantial retaining wall of concrete. The Shakspeare Tunnel (which forms the portion of the same line nearest to Dover) had been for some time completed, and its mouth or entrance from Folkestone is about 400 feet eastward of the central part of the base of Round Down. This intervening point, of about 300 feet in length, was therefore to be passed either by a tunnel or an open cutting: the former had been commenced, and the preliminary driftway formed on the upper level, when heavy slips occurring on either side, which partially extended to the cliff in question, materially impaired its stability: the

---

kiln till it assumes a red colour. Subject to the severe test of Brard's process (boiling in sulphate of soda), slate is found to betray no symptoms of decomposition.

* Account of the Demolition and Removal by Blasting of a portion of the Round Down Cliff, near Dover, in January, 1843. By Lieut. Hutchinson, R.E. Prof. Papers of Royal Engineers, vol. vi. p. 188.—(*Weale.*)

idea of tunnelling was therefore abandoned, and it was determined to remove this point by blasting, and to continue the open railway and sea wall up to the mouth of the Shakspeare Tunnel. For this purpose the principle adopted was to employ large charges, placed a little above the intended level of the railway, which, by blowing out the base of the cliff in their front, would cause the downfall of the superincumbent mass. The result fully justified the adoption of this method.

The experience gained in the many and extensive mining operations on this line of railway led to the adoption of a charge in lbs., bearing the proportion of $\frac{1}{32}$ of the cube of the line of least resistance in feet, corresponding with the rule laid down by Sir J. F. Burgoyne, and which had been found sufficient for the works at Dover for blasts with charges varying from 500 to 1200 lbs. This calculation was therefore adopted in the present instance, and, as the length of face to be removed was great (300 feet) three charges were used, one in the centre at the back of the salient point, the others at the distance of 70 feet on either side. From careful measurements the lines of least resistance were found to be—

    Centre . . .   72 feet to face of cliff.
    East and west . .   56 ditto   ditto.

By calculation the quantity of powder required for the centre chamber was found to be 6750 lbs., which was increased to 7500, to provide against contingencies. For each of the two end chambers an allowance of 5500 lbs. was made, and thus the total quantity of powder became 7500 + 5500 + 5500 = 18,500 lbs., or about 8¼ tons. The preliminary driftway already mentioned was used for excavating to the chambers, being about 4 feet wide and 5 feet 6 inches high; but, being cut on the upper level of the intended tunnel, it became necessary to sink shafts a depth of 17 feet to reach the lower level of the chambers 3 feet above that of the railway. These shafts were formed as truncated cones, 5 feet diameter at bottom, and 3 feet at top, thus affording additional security for the retention of the tamping. At the bottom of these shafts, galleries 5 feet 6 inches high were cut at right angles to the driftway, and continued to the points determined for the chambers. These branches also were made of a dove-tailed form, 2 feet wide at the shaft, and 4 feet 6 inches at the other extremity, to assist in securing the tamping. The chambers were excavated at right angles to the galleries, and of the following dimensions, viz.:—

Centre chamber . 13 ft. by 5 ft. 6 in. high, by 4 ft. 6 in.
Two ends of do. . 10 ft. by 5 ft. 6 in. high, by 4 ft.

The voltaic batteries consisted of fifty-four of Daniell's cylinders, being eighteen for each mine. These cylinders were made of sheet copper, 32 oz. to the superficial foot, 1 foot 10 inches long, and $3\frac{1}{2}$ inches diameter, placed in sets of six each, each set in a deal box 1 foot 4 inches by 1 foot, by 1 foot 9 inches high, with a lid 4 inches deep. The zinc rods were 1 foot 8 inches long, and $\frac{3}{4}$ inch diameter. These batteries differ from the common plate-battery also in employing two liquids separated by a partition of animal membrane, or plaster of Paris. The liquid into which the zinc rod is plunged is sulphuric acid diluted with eight times its quantity of water. The liquid on the outside of the partition is a saturated solution of sulphate of copper.

Besides the eighteen Daniell's cylinders, two plate-batteries were applied for each mine in order to guard against the injury which it appeared Daniell's apparatus was liable to by the low temperature of the season. In these plate-batteries the solution was weaker, being only 1 of the acid to 12 water. The plate-batteries were contained in boxes 3 feet 2 inches by 12 inches and 8 inches deep, of 1 inch deal, and each divided into 20 cells $1\frac{1}{4}$ inch wide, by partitions of $\frac{3}{8}$ inch deal, the cells being coated with a waterproof composition of spirits of wine and sealing-wax. The 20 pairs of plates forming the series, each composed of a zinc plate $\frac{3}{8}$ inch thick, in a rectangular case of copper 10 inches long, $1\frac{1}{8}$ inch wide, and 8 inches deep, without top or bottom, were united by stout copper wire attached to the zinc plate of one pair and the copper plate of the adjoining pair.

For the purpose of firing, it was determined to use three sets of wires and three separate batteries. The total length of wire required was thus 6000 feet, being 2000 for each mine, and $\frac{1}{8}$ inch diameter. The wires were coated with a composition of pitch 8 to bees'-wax 1, and tallow 1, and covered with a coarse cotton tape, bound round while the composition was hot. The two wires thus prepared were insulated by being laid one on each side of a $1\frac{1}{4}$ inch rope, and overlaid by a stout packthread turned once round each wire at every coil. The whole was then bound over with 2 yarn-spun yarn, and finally coated with the same composition of pitch, wax, and tallow. One man with a labourer prepared this wire at the rate of 6 feet per hour.

The cartridges or bursting charges were made 9 inches long and 2 inches in diameter, the top and bottom being secured by cork bungs. The priming wires, $\frac{1}{12}$ of an inch in diameter, projected 1 foot 6 inches beyond the top. They were passed through a slip of wood $1\frac{3}{4}$ inch square and $\frac{1}{4}$ inch thick, inserted at the centre of the tube, and their ends clenched on a circular plate $1\frac{3}{4}$ inches diameter, lying over the lower cork bung. The platinum wire was fixed half-way between the two. The powder, which was of the finest sporting quality, was poured in from the top. The upper bung, being in halves, one-half was removed for this purpose, and refixed when the tube was filled: the top and bottom were then coated with a waterproof composition. Two of these cartridges, after being tested by the galvanometer, were attached to each of the conducting wires, the wires being perfectly insulated, and covered with spun yarn. The cartridges were extended to a distance of 3 feet apart, and buried in the centre tier of loose powder in the box. The wires were then carried from the chambers along grooves cut in the sides of the galleries, shafts, and driftway, and covered with wooden troughs.

For the tamping, a dry wall was built in chalk across the end of each chamber, and the galleries and shafts were afterwards filled in with the same material, which was extended in the driftway to 10 feet on each side of the top of the shafts. The tamping occupied about twenty hours with four men to each mine.

For containing the charges of powder, the deal cases were constructed within the chambers of the following dimensions:—

Boxes.—Centre 11 ft. by 4 ft. 3 in. by 3 ft. 6 in.
   ,,   Ends 10 ft. by 2 ft. 8 in. by 3 ft. 3 in.

The powder was deposited in the boxes, in bags holding 50 lbs. each; the tiers above the centre were placed with their mouths downwards, those below it with their mouths upwards.

The circuit being completed at each battery by pre-arranged words of command, the mines were fired simultaneously on the 26th January, 1843, at 2·20 P.M. The ignition was followed by a deep hollow sound, and the effect was described to resemble the appearance of the fall, or rather flowing, of lava from the side of a mountain, assuming a gently undulating wave-like motion, as the mass slid along the bottom and into the sea. The mass fell exactly as had been desired, and the quantity removed was estimated at more than 400,000 cubic yards, being a parallel mass or slice from the inclined face of the cliff, and averaging 380 feet in height by 80 feet in thickness, and 360 feet in length of face. The operation thus successfully conducted and completed, without the slightest accident, saved about 7000l. to the South-Eastern Railway Company, and reflects the highest credit upon Lieut. Hutchinson, Messrs. Cubitt, Wright, and Hodges, and all other parties concerned.

PART II.

STONES USED IN BUILDING.

To the architect and the engineer, alike, is a knowledge of the materials which he has to employ, or to deal with, of supreme importance. And, were it worth while to inquire the rank which each class of materials occupies in the scale of professional consideration, we should start upon the fact that those substances which, for their substantial properties and enduring character, are selected as the foundations and the exposed external parts of our buildings and structures, claim our first and most attentive examination. Frequently, indeed, local circumstances will define the limits within which the selection must be made; but there are few, if there are any, cases in which the designer is altogether denied the privilege of selection; while in many instances an acquaintance with the materials at his disposal will open a wide field for the exercise of his judgment in aiming at current economy and permanent preservation.

In the few pages left to us in the present volume, we propose to follow up the account given of the methods of getting or quarrying stones, &c., by describing the several properties of the leading classes of those materials which fit them for the purposes of construction, and which should lead to the adoption of one or other of them, according to the service they may be required for.

Without attempting to observe the classification laid down by geologists, we propose to treat of stones under the four general divisions of *Granites, Slates, Limestones,* and *Sandstones,* including under each of these general heads such constituents or subordinate combinations of each as may appear essential to be noticed in this rudimentary outline of a really extensive and highly important department of practical science.

GRANITE.

This rock, which appears to have originally been a fused mass, and subsequently to have undergone the process of crystallisation, is of a *granular* structure, that is, consisting of separate grains of different substances, united, apparently, without the aid of any intermediate matter or cement. These substances are *quartz, felspar,* and *mica,* each of these being a compound. The infinite variety of proportions in which these several constituent elements are united in the mass, occasions the great diversities of colour and appearance of the several kinds of granite, and also affects in a much more important manner the enduring characteristics of this valuable material. Thus its colour varies from light grey to a dark tint closely approaching black, and is to be found of all shades of red, and many green. Of the constituents of granite, *quartz* is a substance of a glassy appearance, and of a grey colour, and is composed of a metallic base *silicium* and *oxygen: felspar* is also a crystalline substance, but commonly opaque, of a yellowish or pink colour, composed of silicious and aluminous matter, with a small proportion of lime and potash; *mica,* a glittering substance, principally consists of clay and flint, with a little magnesia and oxide of iron. Instead of the mica, another substance, called *hornblende,* is found in some granites: hornblende is a dark crystalline substance, composed of flint, alumina, and magnesia, besides a large proportion of the black oxide of iron. Granites in which hornblende exists are sometimes called Syenite, having been first found in the island of Syene, in Egypt. A similar rock is found in Scotland, in Aberdeen, and in the Isle of Arran.

Granite is found in mountain-chains, and usually in rugged outlines, in nearly all parts of Europe. In the Pyrenees it is found in masses of piles or columns. In Germany it forms the material of those famed scenes of supernatural tradition, the Brocken and the Hartz Mountains. Granite forms the bold ridges of Switzerland and the Savoy. In England, this valuable rock occurs in Devonshire and

Cornwall, also in North Wales, Anglesea, the Malvern Hills in Worcestershire, Charnwood Forest in Leicestershire, and in Cumberland and Westmorland. Granite rises near the foot of Skiddaw in Cumberland, and there are masses of it on the banks of Ulswater. It is found in the mountains of Armagh and Wicklow in Ireland; and Mr. Bakewell supposes that the same formation of this rock, which occurs on the western side of England, is continued under the Irish Channel, or, if interrupted, it rises again in the Isle of Man, and in the counties of Dublin and Wicklow in Ireland. Granite blocks are found in the beds of some of the rivers in the north-west parts of Yorkshire, and in clay pits in Cheshire and Lancashire, at great distances from any quarries where the stone is obtainable. Most of the granite met with in Charnwood Forest is of the kind called *Sienite*, and already described. A continuation of the same description of rock appears upon the surface, near Bedworth, in Warwickshire.

In the granite of Cornwall the felspar is usually white, but in the Scotch granite it is commonly red. Among the celebrated works of which granite forms the material, may be named the monuments of Ancient Egypt and of Thebes, the celebrated block used as a pedestal for the statue of Peter the Great in St. Petersburgh, the columns in the church of the Casan in the same city; and in England the columns in the King's library in the Museum, and the bridges of London, &c., &c.

Although all granites are similar in structure, the difference in the proportions of its constituent substances occasions great difference in its enduring and useful properties. Some varieties are exceedingly friable, and liable to decomposition, while others, including that known as Sienite, suffer but imperceptibly from moisture and the atmosphere. Some remarks upon the properties of a Sienite, the "*Herm Granite*" of Guernsey, published in the first volume of the Transactions of the Institution of Civil Engineers, may be here usefully quoted:—

"The Herm granite, as compared with Peterhead, and Moorstone from Devon or Cornwall, is a highly crystallised intermixture of felspar, quartz, and hornblende, with a small quantity of black mica, the first of these ingredients hard and sometimes transparent in a greater degree than that found in other British granites,—the contact of the other substances perfect. It resists the effect of exposure to air, and does not easily disintegrate from the mass when mica does not prevail; but, as this last is usually scarce in Guernsey granites, the mass is not deteriorated by its presence as in the Brittany granites, where it abounds, decomposes, stains, and pervades the felspar, and finally destroys the adhesion of the component parts:—*vide* the interior columns of St. Peter's Port church (Guernsey), which is built of it, for an instance. The quartz is in a smaller quantity, and somewhat darker than the felspar in colour: the grains are not large, but uniformly mixed with the other ingredients. The hornblende, which appears to supply the place of mica, is hard and crystallised in small prisms, rarely accompanied by chlorite: its dark colour gives the grayish tone to this granite, or when abundant forms the *blue* granite of the Vale parish. This substance is essentially superior to mica in the formation and durability of granites for strength and resistance; consequently its presence occasions more labour in working or facing the block, and its specific gravity is increased. The mica is inferior in quantity to the hornblende, and usually dispersed in small flakes in the mass;—it may, with chlorite, be considered rare."

"The compact nature of a close-grained granite, such as the Vale and Herm stone, having the felspar highly crystallised and free from stained cracks, seems well calculated to resist the effect of air and water. When the exterior *bruised* surface of a block has been blown off, it is better disposed to resist decay;—if the surface blocks of the island are now examined after the lapse of ages, it will be found to have resisted the gradual disintegration of time in a superior degree, when compared with *large grained* or *porphyritic* granite; when exposed to water and air, there is no change beyond the polish resulting from *friction* of the elements. Among the symptoms of decay, disintegration prevails generally among granites, usually commencing with the decomposition of the mica; its exfoliating deranges the cohesion of the grains, and it may be considered then to be the more frequent mode of decay."

"The churches of the Vale and St. Sampson (Guernsey), although much of the materials are French and Alderney, bear many proofs of the above remarks: these erections date A.D. 1100—1150. The ancient buildings of decided Herm and Vale stone must be sought for among the old houses in the northern parishes of Guernsey, where they not only encounter the effect of air and water (rain), but the sea air and burning rays of the sun. Disintegration alone appears going on by slow degrees, but in no case affecting the interior of the stone, and so gradual and general as not to deface the building materially; indeed, the oldest proofs taken from door-posts, lintels, and arches, have scarcely lost their original sharpness or sculpture. The pier of St. Peter's Port and bridge of St. Sampson's may also be mentioned.

"The shore rocks in like manner do not show any material change of surface by wearing; where the force of the tide is strongest, a slight smoothness alone may be observed on the exterior, and in many instances each substance possesses this polish *without being levelled down to a face.*"

"Vale stone on the northern point of Guernsey produces a finer grained Sienite than Herm, more hornblende in it, and specific gravity greater. The Herm is somewhat larger grained, but equally good for every erection where durability is the chief point. The *Cat-au-roque* stone in the western part of Guernsey must be considered of a different structure to the above: it is a fair and good stone, and appears to last well; its schistose texture must ally it to the gneiss series. In colour it is much the same as the blue granites, the felspar is brilliant and the hornblende prisms are well defined; there is more chlorite in it, and it is easier to work."

From the same volume, the following interesting results of experiments upon the relative wear of seven different kinds of granite and one whinstone are extracted, as peculiarly valuable in the choice of materials which will be exposed to similarly destructive forces.

"TABLE *showing the result of Experiments made under the direction of Mr. Walker, on the Wear of different Stones in the Tramway on the Commercial Road, London, from 27th March,* 1830, *to* 24*th August,* 1831, *being a period of seventeen months.*

Name of stone.	Sup. area in feet.	Original weight.			Loss of weight by wear.	Loss per sup. foot.	Relative losses.
		cwt.	qrs.	lbs.	lbs.	lbs.	
Guernsey	4·734	7	1	12·75	4·50	0·951	1·000
Herm	5·250	7	3	24·25	5·50	1·048	1·102
Budle	6·336	9	0	50·75	7·75	1·223	1·286
Peterhead (blue) ...	3·484	4	1	7·50	6·25	1·795	1·887
Heytor	4·313	6	0	15·25	8·25	1·915	2·014
Aberdeen (red)	5·375	7	2	11·50	11·50	2·139	2·249
Dartmoor	4·500	6	2	25·00	12·50	2·778	2·921
Aberdeen (blue)......	4·823	6	2	16·00	14·75	3·058	3·216

STONES USED IN BUILDING.

"The Commercial Road stoneway, on which these experiments were made, consists of two parallel lines of rectangular tramstones, 18 inches wide by a foot deep, and jointed to each other endwise, for the wheels to travel on, with a common street pavement between for the horses. The tramstones subjected to experiment were laid in the gateway of the Limehouse turnpike, so as of necessity to be exposed to all the heavy traffic *from* the East and West India Docks. A similar set of experiments had previously been made in the same place, but for a shorter period (little more than four months), with, however, not very different results, as the following figures, corresponding with the column of "*relative losses*," in the foregoing table will show.

"Guernsey	. .	1·000	Peterhead (blue)	1·715
Budle	. . .	1·040	Aberdeen (red) .	2·413
Herm	. . .	1·156	Aberdeen (blue)	2·821

All the above stones are granites except the Budle, which is a species of whin from Northumberland, and they were all new pieces in each series of experiments."

The relative hardness, or resistance to crushing, of the several kinds of granite in common use, is exhibited in the following Table of Experiments made with an hydrostatic press on specimens of the sizes stated in the Table.

These experiments were made with a 12-inch press, the pump one inch diameter, and the lever 10 to 1;—the mechanical advantage therefore $144 \times 10 = 1440$ to 1. The weights on the lever were added by 7 lbs. at a time;—each addition therefore equivalent to $1440 \times 7 = 10{,}080$ lbs. or $4\frac{1}{2}$ tons.

In consequence of the smallness of the specimens, the press was filled with blocks to the required height, and with these the surplus effect of the lever was $4\frac{1}{2}$ lbs. at 10 to 1, which strictly should be added to the pressure; but, as the friction of the apparatus is equal to the effect of the lever, it is dispensed with in the calculation.

The column containing the pressure per square inch required to produce a fracture gives the true value of the stone, as the weight that does so would possibly completely destroy the stone if allowed to remain on for a length of time. It should also be observed that, from the exceedingly short time allowed for the experiments, the results are probably too high.

Table of Experiments on Granite.

Description of Stone.	Weight of each specimen.	Dimensions.	Surface exposed to pressure.	Pressure required to fracture stone.			Pressure required to crush stone.		
				Total to each specimen.	Per sup. inch of surface.	Average per sup. inch.	Total to each specimen.	Per sup. inch of surface.	Average per sup. inch.
	lbs. oz.	Lineal Inches.	Sup. in.	Tons.	Tons.	Tons.	Tons.	Tons.	Tons.
Herm	6 6 6 6	4 ×4 ×4 4 ×4 ×4	16 16	80·0 72·5	5·00 4·53	4·77	116·0 96·4	7·25 6·03	6·64
Aberdeen (blue) . .	5 0 5 1½	4 ×4¼ ×3 4 ×4½ ×3	17 18	81·0 63·0	4·76 3·50	4·13	85·5 76·5	5·03 4·25	4·64
Heytor	4 7 4 8	4 ×4 ×3 4 ×4 ×3	16 16	67·5 58·5	4·22 3·66	3·94	103·5 94·5	6·47 5·91	6·19
Dartmoor . . .	4 10 4 8	4 ×4 ×3 4 ×4 ×3	16 16	67·5 45·0	4·22 2·81	3·52	103·5 72·0	6·47 4·50	5·48
Peterhead (red) . .	5 5 4 12	4½ ×4 ×3⅜ 4½ ×4 ×3	18 18	58·5 45·0	3·25 2·50	2·88	94·5 81·0	5·25 4·50	4·88
Peterhead (blue-grey)	5 3½ 5 4	4½ ×4½ ×3⅜ 4½ ×3⅜ ×3⅜	18·6 17·5	58·5 45·0	3·14 2·57	2·86	85·5 72·0	4·60 4·11	4·36
Penryn. . . .	5 7 5 4	4½ ×4 ×3 4½ ×4 ×3½	18·5 18	63·0 31·5	3·41 1·75	2·58	72·0 54·0	3·90 3·00	3·45

The method pursued in quarrying the Peterhead granite has been already described at page 78.

Some experiments upon the granites of Ireland were reported to the Geological Society of Dublin, in January, 1844. These experiments were instituted to show (among other results) the weight of water which the stones would imbibe when immersed in water. The size of the stones immersed was 14 inches long and 3 inches square. They were placed on their ends in 16 inches of water, and were uniformly immersed for 88 hours, having been brought to a dry state before immersion by being kept some time in a room at the ordinary temperature of domestic apartments. They were carefully weighed before and after the immersion. The average weight of several specimens of the stones was 170 lbs. per cubic foot: the maximum being 176, and the minimum 143 lbs. The Newry and Kingston granites absorbed a quarter of a pound of water to the cubic foot; the Carlow, from 1½ to 2 lbs.; and the Glenties, from Donegal (between granite and gneiss), 4 lbs.

Serpentine and *Porphyry* are sometimes classed as varieties of granite; but more properly as distinct rocks of the same primary character as granite. Serpentine is a valuable material for the ornamental purposes of architecture, being distinguished by its variety and richness of colours; these are generally light and dark green of various shades, intermixed in spots or clouds, resembling the spots on the skins of serpents (whence its name). Some varieties of this rock have also a red colour; others are found having an intermixture of crystalline white marble, and these are known as *verde antique*, highly valued for sculpture of ornamental character. Some kinds of serpentine are crystalline, and are called *schiller spar* or *diallage*. In some of these rocks, magnesia exists in the proportion of 48 per cent.

Serpentine is found in Cornwall with a micaceous rock overlying the granite, and forms part of the promontory called the Lizard Point; it also occurs in the same county, near Liskeard. It is not met with elsewhere in England; but in Anglesea beautiful varieties of the red and green-coloured are found in beds of considerable thickness, associated with the common slate rocks of the district. The mixture of serpentine with *talc* or *steatite* becomes soft, and forms the substance called *potstone*, which resists fire, and is used in Switzerland, Lombardy, and Egypt, for culinary and other purposes.

SLATE.

In treating of slate as a material of common utility to the builder, it will be scarcely necessary to do more than enumerate the several varieties which are known under the general term slate, but distinguished one from another, and from the slate of general utility, by a peculiar prefix. Thus, the geologist recognises *mica slate, talcous slate, flinty slate*, and *common*, or *clay slate*. Of these, the last only is a material of extended use in the arts of building and construction. Of the others, the first, or mica slate, is a compound of quartz and mica, with sometimes a little felspar. The several varieties of this rock are dependent on the proportion of mica contained in each, and on the comparative fineness or coarseness of the constituent particles. If the felspar abound, the compound passes into a form of rock known as *gneiss;* and if the mica exist in only small proportion, the material assumes the form of *quartz rock*. The talcous slates are distinguished by their green colour, and contain a large proportion of magnesia. One variety of this kind of slate contains particles of quartz, and is sometimes used for hones, under the name of *whetstone* slate. The third variety on our list, flinty slate, contains a larger proportion of flinty or silicious matter, and assimilates to the scaly structure of flint. Possessing the silicious earth, but *not* the scaly kind of formation, this slate passes into a rock known as *hornstone* by us, and as *petro-silex* by the French. If it contains felspar in crystals, it is distinguished as *hornstone porphyry*. All of these varieties of slate are found alternating with each other in the same rocks in North Wales, and in the Charnwood Forest, Leicestershire. The common or clay slate abounds in the most rocky districts, and is found lying upon granite, gneiss, or mica slate. The several varieties of character—from the extreme crystalline to the extreme earthy—are found occupying positions in regular order, from the primary towards the transition rocks. Clay slate, as its name implies, consists chiefly of clay in an indurated condition, and occasionally containing particles of mica and quartz; and, in some of the coarser kinds, grains of felspar and other fragments of the primary rocks. In the extreme admixture of these foreign substances, clay-slate approaches the nature of the rock known as greywacke. The beds of clay-slate are invariably stratified, the thickness of the strata, however, varying from a fraction of an inch to many feet.

Its laminar texture admits of a ready separation into thin plates, and thus endows it with a supreme value for roofing and other purposes, in which great density and comparative impermeability are required to coexist with a minimum thickness and weight. In our preceding pages, we have quoted a highly interesting description of the method pursued in quarrying this invaluable material. We may here add that, besides the immense quarries which are worked, as there described, at Bangor, in North Wales, slate is also procured from the counties of Westmoreland, York, Leicester, Cornwall, and Devon, and that Scotland is supplied with this useful material from Balahulish and Easdale.

In Ireland slate quarries are worked to a considerable extent at Killaloe, Valentia, and in Wicklow County; the slates from the latter of which are said to resemble those from Bangor in quality and extreme tenacity, and can be raised in blocks 30 feet long, 4 or 5 feet wide, and from 6 to 12 inches in thickness. The weight of slates varies from 174 to 179 lb. per cubic foot; the average weight of several specimens being 177 lb. When immersed in water in the manner already described for granite (page 95), the clay-roofing slate absorbed less than a quarter of a pound of water; while a softer quality of clay slate, from the neighbourhood of Bantry, absorbed nearly 2 lb.

SANDSTONES.

These rocks, belonging, geologically, to various positions in the order of the strata of which the exterior of the earth is composed, and sometimes alternating with the variety of limestones, are widely distributed, or rather frequently met with, superficially, in exploring the surface of our island.

The vast accumulation of beds, known as forming the *Silurian System*, and of which sandstone is the principal member, is found to extend, with some interruptions, from Carmarthenshire and Radnorshire, in the west, to Dudley in Staffordshire, in the central part of England. At the western boundaries of this system the sandstone is known as Llandillo flagstones; it occurs also at Caer-Caradock in Shropshire, and is there and elsewhere known as Caradock sandstone. Some varieties of this stone are found to contain a large proportion of lime, occasionally in the form of shells, and are hence sometimes called shelly limestones and sandy limestones.

Sandstones are principally silicious, and possess various degrees of induration. The chemical analysis of several specimens of sandstone from four quarries, viz., Craigleith, in Edinburghshire;—Darley Dale, in Derbyshire; Heddon, in Northumberland, and Kenton, in the same county, gives their average constituents thus:—

Silica	95·725
Carbonate of lime	1·065
Iron alumina	2·150
Water and loss	1·060
	100·000

The average weight of these four kinds of stone was found to be 142 lb. 7 oz. per cubic foot. Adopting the same high authority from which these analyses are derived, viz., the scientific Commission appointed to visit quarries and examine the qualities of stones to be used in building the New Houses of Parliament, the composition of these four kinds of sandstone appears to be as follows:—*Craigleith:* Fine quartz grains with a silicious cement, slightly calcareous; occasional plates of mica; colour, whitish gray.—*Darley Dale:* Quartz grains of moderate size and decomposed felspar, with an argillo-silicious cement; ferruginous spots, and plates of mica; colour, light ferruginous brown.—*Heddon:* Coarse quartz grains, and decomposed felspar, with an argillo-silicious cement; ferruginous spots; colour, light brown ochre.—*Kenton:* Fine quartz grains with an argillo-silicious and ferruginous cement; mica in planes of beds; colour, light ferruginous brown.

From the nature of the composition of sandstones, (of which the above four are described as average specimens,) it results that their resistance against, or yielding to, the decomposing effects to which they are subjected, depends to a great extent, if not wholly, upon the nature of the cementing substance by which the grains are united; these latter being comparatively indestructible. From the nature of their formation, sandstones are usually laminated, and more especially so when mica is present, the plates of which are generally arranged in planes parallel to their beds. Stones of this description should be carefully placed in constructions, so that these planes of lamination may be horizontal, for if placed vertically, the action of decomposition will occur in flakes, according to the thickness of the laminæ. Indeed the best way of using all descriptions of stone is in the same position which they had in the quarry, but this becomes a really imperative rule with those of laminated structure.

Uniformity of colour is a tolerably correct criterion of uniformity of structure, and this constitutes, other circumstances being equal, one of the practical excellencies of building stones. The great injury occasioned to these materials by their absorption of moisture, leads properly to a preference for such stones as resist its introduction, for all above-ground purposes. Those which imbibe and retain moisture are especially liable to disruption by frost if exposed. The simplest method of ascertaining the disposition of a stone to imbibe moisture is to immerse it for a lengthened period in water, and to compare the weight of it before and after such immersion.

A method recently introduced of determining the susceptibility of a stone to injury by frost, is to dip the stone in a solution of some salt, and then suspend it for several days over the solution, repeating the process several times, so as to allow the salt to crystallise on the surface of the stone. If the stone can resist frost the solution will remain free from sand or fragments of stone, but if otherwise, the edges of the stone will be found deposited in the vessel beneath.

As examples of the durability of sandstone in buildings, the following may be instanced :—Ecclestone Abbey, built in the thirteenth century, near Barnard Castle, Durham, in which the minute ornaments and mouldings remain still in excellent condition. The circular keep of Barnard Castle (14th century), is also in fine preservation. Tintern Abbey (13th century), built of red and grey sandstones of the vicinity, is for the most part in perfect condition. Whitby Abbey (13th century), is built of stone similar to that in the vicinity, and is generally in good condition, excepting the west front, which is very much decomposed. The stone used is of two colours, brown and white, of which the latter is uniformly in the best preservation. The enrichments on the east front are all in good condition. Of Ripon Cathedral, the west front, the transepts and tower, were built in the 12th and 13th centuries of a coarse sandstone of the vicinity, and remain in very fair condition. Rivaulx Abbey, built in the 12th century, of a sandstone one mile from the ruins, is generally in excellent condition. The west front is slightly decomposed, but the south front is remarkably perfect, even to the preservation of the original tool-marks.

Experiments upon several sandstones to ascertain the force required to crush them, performed with the hydrostatic press, as already described of the specimens of granite (page 93) gave the following results—

Table of Experiments upon several Sandstones.

Description of Stone.	Weight of each specimen.	Dimensions.	Surface exposed to pressure.	Pressure required to fracture stone.			Pressure required to crush stone.		
				Total to each specimen.	Per sup. inch of surface.	Average per sup. inch.	Total to each specimen.	Per sup. inch of surface.	Average per sup. inch.
	lb. oz.	Lineal Inches.	Sup. In.	Tons.	Tons.	Tons.	Tons.	Tons.	Tons.
Yorkshire (Cromwell bottom)	12 8 12 5	5⅜ × 5 × 5¼ 5½ × 5 × 5⅜	27·5 27·5	81·0 76·5	2·95 2·78	2·87	121·5 95·5	4·42 3·47	3·94
Craigleith	11 10 11 6	5 × 5 × 5¼ 5 × 5 × 5⅜	25 25	63·0 31·5	2·52 1·26	1·89	85·5 63·0	3·42 2·52	2·97
Humbie	17 10 17 3	6 × 6 × 6 6 × 6 × 6	36 36	72·0 49·5	2·00 1·37	1·69	81·0 67·5	2·25 1·87	2·06
Whitby	16 10 15 12	6 × 6 × 6 6 × 6 × 6	36 36	36·0 36·0	1·00 1·00	1·00	40·5 36·0	1·12 1·00	1·06

LIMESTONES.

The class of limestones, including the magnesian limestones and the oolites, is one of extreme importance in the building arts, comprehending some of the most advantageous materials of construction, and combining great comparative durability with peculiar facilities for working, in which they surpass the sandstones; as those do the rocks of a primary and unstratified character. Of the limestones and the oolites, the principal material is carbonate of lime. The magnesian limestones contain a quantity of carbonate of magnesia, in some cases nearly equal to that of the carbonate of lime. The analyses of four varieties of oolites and three limestones are given below. The oolites were from Ancaster in Lincolnshire; from Bath; from Portland, Dorsetshire; and from Ketton in Rutlandshire. The three limestones were from Barnack, Northamptonshire; Chilmark, Wiltshire, and Ham Hill in Somersetshire.

	OOLITES.			
	Ancaster.	Bath.	Portland.	Ketton.
Silica	...	...	1·20	...
Carbonate of lime	93·59	94·52	95·16	92·17
Carbonate of magnesia	2·90	2·50	1·20	4·10
Iron alumina	·80	1·20	·50	·90
Water and loss	2·71	1·78	1·94	2·83

	LIMESTONES.		
	Barnac.	Chilmark.	Ham Hill.
Silica	...	10·4	4·7
Carbonate of lime	93·4	79·0	79·3
Carbonate of magnesia	3·8	3·7	5·2
Iron alumina	1·3	2·0	8·3
Water and loss	1·5	4·2	2·5

All of these stones contain a slight trace of bitumen in their composition. The weights of the several kinds are as follow:—Of the oolites, the Ancaster stone, 139 lb. 4 oz. per cubic foot.—Bath, 123 lb.—Portland, top bed, 135 lb., best or lower bed, 147 lb., mean, say 141 lb.—Ketton; 128 lb. 5 oz. Of the three limestones, the weights are, Barnack, 136 pounds, 12 oz.—Chilmark, 153 lb. 7 oz.; and Ham Hill, 141 lb. 12 oz. The composition of these stones is described as follows:—First, Oolites.—*Ancaster:* fine oolitic grains, cemented by compact and often crystalline carbonate of lime; colour, cream.—*Bath:* chiefly carbonate of lime in moderately fine oolitic grains, with fragments of shells. The weather bed of the quarry is generally used for plinths, strings, cornices, etc.; the corn grit for dressings; the scallet, which is the finest in grain, is used for ashlar. Eight quarries are opened on the Box escarpment, many of them of great antiquity. The colour is that of cream.—*Portland:* oolitic carbonate of lime, with fragments of shells; colour, whitish brown.—*Ketton:* oolitic grains of moderate size, slightly cemented by carbonate of lime; colour, dark cream. Secondly, Limestones.—*Barnack:* carbonate of lime, compact and oolitic, with shells, often in fragments; coarsely laminated in planes of beds; colour, light whitish brown. This stone is used for troughs and cisterns, and is perfectly impervious.—*Chilmark:* carbonate of lime, with a moderate proportion of silica, and occasional grains of silicate of iron; colour, light greenish brown.—*Ham Hill:* compact carbonate of lime with shells, chiefly in fragments, coarsely laminated in planes of beds; colour, deep ferruginous brown.

Of four varieties of magnesian limestone, the following are the analyses:—

	MAGNESIAN LIMESTONES.			
	Bolsover.	Huddlestone.	Roach Abbey.	Park Nook.
Silica	3·6	2·53	·8	...
Carbonate of lime . .	51·1	54·19	57·5	55·7
Carbonate of magnesia	40·2	41·37	39·4	41·6
Iron alumina	1·8	·30	·7	·4
Water and loss . . .	3·3	1·61	1·6	2·3

The weights of these limestones were as follows:— Bolsover, 151 lb. 11 oz. per cubic foot; Huddlestone, 137 lb. 13 oz.; Roach Abbey, 139 lb. 2 oz.; and Park Nook, 137 lb. 3 oz. The composition of them is as follows:

—*Bolsover:* chiefly carbonate of lime and carbonate of magnesia, semi-crystalline; colour, light yellowish brown.—*Huddlestone:* similar component parts; colour, whitish cream.—*Roach Abbey:* chiefly carbonate of lime and carbonate of magnesia, with occasional dendritic spots of iron or manganese; semi-crystalline; colour, whitish cream.—*Park Nook:* similar constitution, and of cream colour. Sinks and tanks are made of this stone, but the water wastes in them.

The following structures of magnesian limestones may be distinguished:—Southwell Church, Nottinghamshire, built in the 10th century, and now in perfect condition. Koningsburgh Castle, in Yorkshire (Norman), built of coarse-grained and semi-crystalline magnesian limestone, from the adjacent hill, is in excellent preservation. The mortar has in many parts disappeared, but the edges of the joints remain perfect. The Church at Hemingborough, in Yorkshire (15th century), is in a very perfect state. It is built of a white crystalline magnesian limestone, resembling that from Huddlestone. The entire building is in a perfect state; even the spire, where no traces of decay are apparent.

It is remarked that magnesian limestone appears capable of resisting decomposing action in proportion as its structure is crystalline, and the late Professor Daniell gave it as his opinion, based upon experiments, that "the nearer the magnesian limestones approach to equivalent proportions of carbonate of lime and carbonate of magnesia, the more crystalline and better they are in every respect."

Among the buildings constructed of oolitic and other limestones, the following are deserving of notice:—Byland Abbey, Yorkshire, of the 12th century, built partly of a silicious grit (principally in the interior), and partly (chiefly in the exterior) of a compact oolite, from the Wass quarries in the vicinity. The west front, which is of the oolite, is in perfect condition, even in the dog's-tooth and other florid decorations of the doorways, &c. This building is generally covered with lichens.

Bath.—*Abbey Church* (1576), built in an oolite from the vicinity. The tower is in fair condition, the body of the church in the upper part of the south and west sides much decomposed. The lower parts (formerly in contact with buildings) are in a more perfect state: the relief in the west front of Jacob's ladder are in parts nearly effaced.—*Queen's Square,* north side, and the obelisk in the centre, built about 110 years since, of an oolite with shells, in fair condition. *Circus*, built about 1750, of an oolite in the vicinity, generally in fair condition, except those portions which have a west and southern aspect, where the most exposed parts are decomposed.—*Crescent*, built about 60 years since, of an oolite of the vicinity, generally in fair condition, excepting a few places where the stone appears to be of an inferior quality.

Oxford.—*Cathedral* (Norman, 12th century), chiefly of a shelly oolite, similar to that of Taynton. Norman work in good condition; the later work much decomposed.—*Merton College Chapel* (13th century), of a shelly oolite, resembling Taynton stone, in good condition generally.—*New College Cloisters* (14th century), of a shelly oolite (Taynton) in a good condition. The whole of the colleges, churches, and other public buildings of Oxford, erected within the last three centuries, are of an oolitic limestone from Headington, about 1½ mile from the University, and are all, more or less, in a deplorable state of decomposition. The plinths, string-courses, and such portions of the buildings as are much exposed to the action of the atmosphere, are mostly of a shelly oolite from Taynton, 15 miles from the University, and are universally in good condition.

Saint Paul's Cathedral, London. Finished about 1700. Built of Portland oolite, from the grove quarries on the East Cliff. The building generally in good condition, especially the north and east fronts. The carvings of flowers, &c., are throughout nearly as perfect as when executed. On the south and west fronts large portions of the stone exhibit their natural colour, occasioned by slight decomposition of the surface. The stone in the drum of the dome, and in the cupola above it, appears not so well selected as the rest, yet shows scarcely any decay in those parts.

Westminster Abbey. (15th century). Built of several varieties of stone, similar to Gatton or Reigate, much decomposed, and of Caen stone, generally in bad condition. A considerable portion of the exterior, especially on the north side, has been restored at various periods; nevertheless, abundant symptoms of decay are apparent. The cloisters built of several kinds of stone, are in a very mouldering condition, except where they have been recently restored with Bath and Portland stones. The west towers erected in the beginning of the 18th century, with a shelly variety of Portland oolite, exhibit scarcely any appearance of decay;

Henry the Seventh's chapel, restored about thirty years since, with Coombe Down Bath stone, is already in a state of decomposition.*

In *Spofforth Castle* is a striking example of an unequal decomposition of the magnesian limestone and a sandstone; the former used in the decorated parts, and the latter in ashlar and plain facings. Although the magnesian limestone has been equally exposed with the sandstone, it has remained as perfect as when erected, while the sandstone has suffered considerably. In Chepstow Castle, and in Bristol Cathedral, similar contrasted effects are visible upon the magnesian limestones and sandstones employed.

The late Professor Daniell gives the following valuable summary of the results of his experiments upon the stones we have referred to. "If the stones be divided into classes, according to their chemical composition, it will be found that in all stones of the same class there exists, generally, a close relation between their various physical qualities;—thus it will be observed that the specimen which has the greatest specific gravity possesses the greatest cohesive strength, absorbs the least quantity of water, and disintegrates the least by the process which imitates the effects of weather. A comparison of all the experiments shows this to be the general rule, though it is liable to individual exceptions. But this will not enable us to compare stones of different classes together. The sandstones absorb the least water, but they disintegrate more than the magnesian limestones, which, considering their compactness, absorb a great quantity. The heaviest and most cohesive of the sandstones are the Craigleith and the Park Spring. Among the magnesian limestones that from Bolsover is the heaviest, strongest, and absorbs the least water. Among the oolites, the Ketton Rag is greatly distinguished from all the rest by its great cohesive strength and high specific gravity."

For durability, for crystalline character, combined with a close approach to the equivalent proportions of carbonate of lime and carbonate of magnesia; for uniformity of structure; for facility and economy in conversion; and for advantage of colour, the commissioners declared their preference for the magnesian limestone of Bolsover and its neighbourhood, and, taking into account the extended range and careful precision of their observations, and the skilful manner in which their

* Report of Commissioners.

experiments were designed and conducted, it will be generally admitted that the opinions thus formed and presented to us are excellent guides for our selection of stone in all architectural constructions, and deserve our most careful consideration, if they do not claim our immediate concurrence.

BRIDGES IN SPAIN.

MEMORANDUM ON THE BLOWING UP OF BRIDGES,

BY

GENERAL SIR JOHN F. BURGOYNE, BART.,

OF THE CORPS OF ROYAL ENGINEERS.

DATED DEC. 1814.

BRIDGES IN SPAIN.*

In the destruction of bridges during the Duke of Wellington's compaigns, various methods were adopted according to the circumstances of the case.

The bridges in the Peninsula were usually of stone, the arches from 20 to 40 feet span semicircular, and of one stone of 18 inches or 2 feet in thickness. The loading of the arches was sometimes of solid masonry, but commonly of loose stones or rubbish.

The object required generally was to destroy one arch, and in order to give the enemy the greatest inconvenience and delay, the largest arch and where there was deep water was preferred, excepting when want of time or ammunition made it advisable to select a particular one that might appear weaker than the others.

The simplest principle of mining a bridge was found to be by lodging the powder on the haunch of the arch, and as near as could be on the centre of the width of the bridge, with the line of least resistance through the arch.

The best mode of forming the mine was where the side walls of the bridge above the piers were slightly built and easily got at, and the loading of the arch of loose rubbish;

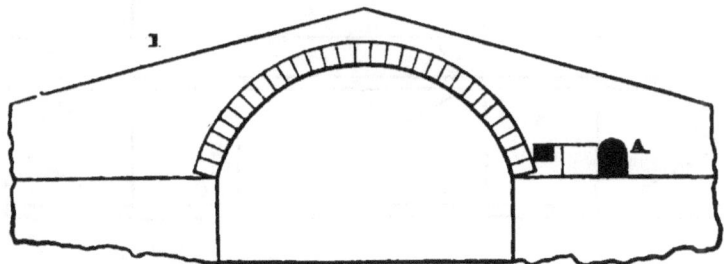

a small gallery was then run in A, fig. 1, about 5 feet from the arch stone; and when at the centre of the width of the

* Extracted from a Treatise on Mines, by J. Landmann, Professor of Fortification and Artillery. Lond. 1815.

bridge, a return was made to the arch, and the powder lodged against it. There are not many occasions where this can be done under a very considerable time; but when practicable it has many advantages: the greatest resistance is obtained to the sides and above; the ammunition is less likely to get injured from wet penetrating to it; there is no obstruction to the road over the bridge while preparing, and less danger of accidents, after it is loaded.

In this case, the powder, saucisson, &c., are applied in the usual manner in mining; and the end to be lighted is kept within the surface of the wall to be sheltered from the weather.

The common and quickest mode of mining a bridge is by sinking down from the road above, to the arch, and lodging the powder in one mass on the centre of its width. To do this with good effect, the shaft C B, fig. 2, should be sunk

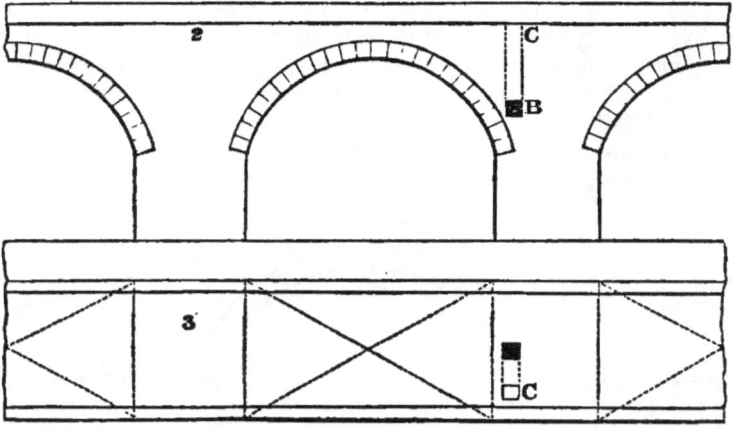

where there will be the greatest resistance gained above and to the sides, as at B. As the arch gives so much more resistance than the materials with which it is loaded, the distance to the surface, therefore, should be twice, or three, or even four times more, at least in those directions, than in that through the arch, in proportion to the nature of those materials.

In this way arches have been blown down with 45 pounds of powder, and after five or six hours of labour.

The shaft should be sunk on one side of the centre of the width of the bridge, as at C, fig. 3, and a little return made at the bottom to gain that situation for the powder, by

BRIDGES IN SPAIN. 111

which means there will be more solid resistance above, and a greater width of road left open during the operation.

In loading, the saucisson was brought up the shaft to within about 1 foot of the surface of the road, and then carried along a gutter or drain to the side of the bridge where it was lighted, whereby the road was entirely cleared, and a premature explosion from accident less likely to occur. The upper surface of the road was drained off as much as possible, to keep the wet from penetrating to the powder.

When there is no time to sink a shaft as deep as might be wished, as great resistance must be obtained as can be, by sinking as deep to the arch as there is time for, and increasing the effect by a loading of as much stone or other heavy materials from the parapet walls or elsewhere as can be applied.

A bridge across the Carrion at Dueñas was required to be mined in great haste, and it was found that the loading between the arches was of solid masonry: an opening was, therefore, made down to the crown of the arch D, figs. 4 and 5, about two feet six inches only; 250 pounds of

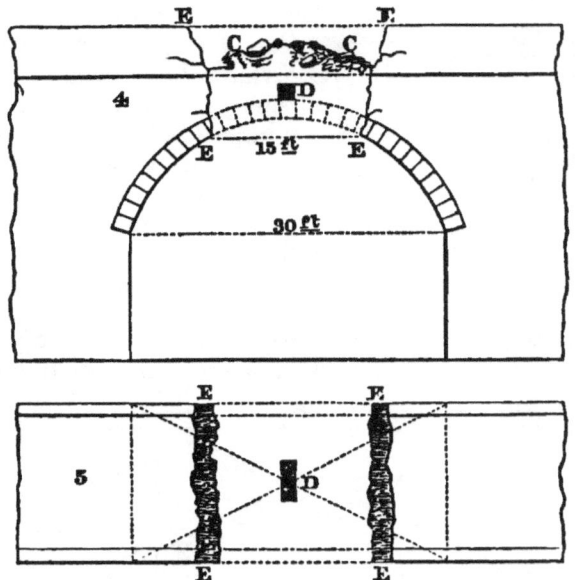

powder were lodged in rather a longitudinal direction along the width of the bridge, and a loading, c c, fig. 4, applied of heavy stones and rubbish, as high above the road of the bridge as could be without preventing carriages from passing; when fired it made a gap, E E E E, across the bridge of fifteen feet, which was about half its span.

The French declare that 100 pounds of powder laid on the crown of an arch, and without loading would destroy it; but, in a strong built bridge, I should be sorry to apply so small a quantity.

As on service, the time at command for this kind of operation is very uncertain; it is a common and good mode to commence preparing in two places, one on the crown of the arch and the other on the haunch, and then if not allowed time sufficient to complete the latter and better mode, the powder can be applied on the crown of the arch, and exploded with or without a loading of rubbish, according to circumstances; and it is much better to do that than to lodge the powder in a shaft only partly sunk down to the haunch, although it should be deeper.

In some cases where the bridge is very wide, and the operation can be carried on with nicety, it may be right to divide the powder into two mines F and G, fig. 6, across its width; but, in a rough operation, I would certainly never divide the powder: for although it was said once that a hole was blown through the centre of a wide arch, and a passage left on each side (which, however, I do not believe), if it was so, certainly that same quantity of powder that gave so nice a shock would not have injured the arch at all if divided.

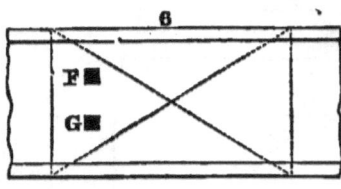

I have seen an instance where about half of the width of a bridge F G H, fig. 7, was blown down, which probably

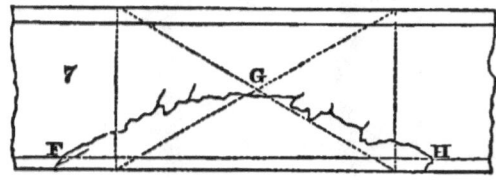

arose from dividing the powder in this manner.

There can be no reason whatever for dividing the powder

between the different sides of the arch, as at I K, fig 8; by doing so a failure took place on the Coruña retreat; and if it succeeds, there can be little doubt but that one of the mines would have done as well. Wherever the powder is divided, the explosion of the whole should be simultaneous; the arrangements require much precision, and the chances of failure are of course multiplied.

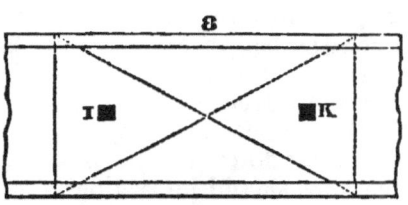

Where a bridge is narrow, there can be no occasion for sinking the shaft down to the arch much deeper than half the width of the bridge, as the want of resistance at the sides will render the additional vertical resistance superfluous. On one occasion a failure occurred from a shaft being sunk down to a pier with the intention of destroying two arches; but which, although great perpendicular resistance was gained, blew out at the sides, and left the arches perfect.

When the effect of a mine can be secured to cut through the arch, the greater resistance that can be given even in that direction the better, as it will increase the effect over the whole width of the bridge.

As it generally happens on service, that the mine cannot be laid according to nice calculation, after applying it in the best way which circumstances will allow, the effect must be gained by increasing the quantity of powder. Under the chance of different difficulties that might occur, it was customary, when practicable, to send two, three, and even four barrels of powder, of ninety pounds each, for the destruction of a bridge, although one would usually be sufficient.

When there was time these mines were loaded with all the precautions commonly used—viz., the powder in a box, and the saucisson in an auget; and when to lay any time, the box and auget were pitched, and covered with straw, tarpauling, &c., to preserve the ammunition dry. When pressed for time and without the proper articles, the powder was lodged in the barrels it was brought in, or laid in a tarpauling, or in bags; and the saucisson was laid without an auget, but with care that the stones or rubbish should not choke it. The mine was lighted by a piece of port-fire tied into the end of the saucisson.

Saucisson is so very easily made and carried, and so advantageous, that latterly we never failed having it with us; in our first mines, indeed, for want of it, we cut off the ends of port-fires diagonally, and tied them together to pieces of stick the length necessary for train; but such a contrivance is very bad, and owing to it Lieutenant Davy was killed on Sir John Moore's retreat, the mine exploding the instant he lighted it, probably from the fire of the composition dropping down to the powder, for which reason the end port-fire should be laid horizontally, and a little clay round it will give additional security.

A small hollow round the powder in a mine will increase its effect.

To destroy wooden bridges powder was sometimes used and applied to the most important supports in the arch, according to its construction; but as there is no other resistance than the air, the quantity of powder should be large; ninety pounds have blown down a strong wooden arch.

The common and best mode with a wooden bridge is to lay the planking bare, and to light a large fire upon them over the piles forming the piers, which will then burn to the water's edge if let alone; but this will not do if the enemy cannot be kept from gaining possession of the bridge for at least twelve hours after the fire is lighted.

ADDENDA.

It may not be considered uninteresting to follow the foregoing subject by adding some account of the signal failures in the blowing up of two bridges by gunpowder, arising from a want of that due consideration and discretion in moments of danger which is so essential and of such paramount importance to the military engineer. The first is that of the failure in destroying and preventing the passage of troops over the bridge between Pesth and Buda during the Hungarian revolutionary war of 1848. The facts were personally communicated to me, after his return to England, by the engineer of the bridge, the late Wm. Tierney Clark, Esq. The second is that of the failure in the destruction of the bridge of S. Martino, near Magenta, during the Italian war of 1858. Of the first I have made some extracts from Mr. Clark's published work of the particulars of the construction of that fine and stupendous work of art:—

"The provisional government sent messages to the directors of the bridge to prepare the approaches, &c., for the passage of troops and cavalry, and afterwards likewise for artillery, and this to be done under the severest penalties. All representations of the danger which might occur from the passage of troops over an unfinished bridge were unavailing, and totally disregarded. This being the case, therefore, the 5-inch longitudinal larch timbers were covered over with cross temporary planking, to save them as much as possible. After this had been done, the bridge was daily crossed by infantry and cavalry regiments, light and heavy artillery, baggage waggons, &c., and the whole of the Hungarian army at last retreated over it, and on the 5th and 6th of January the Imperial troops, to the amount of about 70,000 men, comprising nearly a dozen cavalry regiments, and 270 cannon, passed over it, and took possession of Buda-Pesth."

"After the passage of the Imperial troops on the night of the 29th April over the suspension bridge and bridge of boats, the latter was set fire to and completely destroyed, being previously covered with pitch, tar, and other combustible matter. General Hentzi then caused the whole of the 5-inch platform timbers, as well as those forming the foot-paths, to be cleared away, leaving the cast-iron beams and trussing between the piers standing quite bare. He then caused four large cases, containing about 30 cwt. of gunpowder, to be placed on the beams close to the chains, two on each side, with a train extending out to behind the toll-houses, with orders for the artillerymen to fire it if the Hungarians attempted to force a passage—Pesth being already in possession of the insurgent troops."

"On the night of the 20th of May, the walls of the city were stormed at all points by an overwhelming force; and after a bloody and severe struggle, in which General Hentzi lost his life, the fortress was taken; and on the morning of the 21st, the Hungarian colours appeared floating from the towers. Between 6 and 7 o'clock that morning, and after all was lost, with the exception of the Buda workyard, which was still held by the brave General who commanded them, he—for what reason is not known—set fire to the powder on the bridge with his own hands, blowing himself and about 80 feet of the skeleton of the platform to atoms; after this, all resistance ceased. Five of the beams were blown to pieces, and three were broken, but still hanging. The troops broke into the offices and magazines at the works, and destroyed much valuable property, and it was with the greatest difficulty that the papers and drawings were saved from destruction.

"The bridge from this blasting sustained, nevertheless, but little damage, and the Hungarian General Georgy came and gave orders for the bridge to be repaired for the passage of troops."

"The bridge continued open for the military and a part of the public till the 30th of May, when it was closed, and the works commenced again at all points; and as the authorities stated that the progress should not again be interrupted, a beginning was made to lay the wood pavement on the platform."

"During the war the bridge had been tested by the passage of numerous troops of the different armies of infantry, cavalry, and artillery, with heavy waggons and the usual accompaniments crowding the entire surface of the platform, not only for a day, but for weeks together, day and night, and this when the trussing and bracing, which adds so much to the strength of the platform, was for the most part not erected."

Of the second I have made some extracts from Paper XII., in Vol. 10 of the Papers of the Corps of Royal Engineers, written by General Sir John F. Burgoyne, Bart., G.C.B., 1861.

"The Austrian account of the attempted destruction of the bridge over the Ticino at San Martino, near Magenta and Buffalora, during the campaign of 1859, is of interest, inasmuch as the destruction of this bridge was of the greatest consequence to the Austrians; for when they retreated from it, leaving it in such a state as to be passable for troops, the French were enabled to cross in sufficient numbers to win the battle of Magenta before their opponents were concentrated. Between the Ticino and the town of Magenta, however, they encountered another obstacle, viz., the Grand Canal, which would have checked them for a long time if properly defended; but, although the four bridges across it were mined, only two of them were blown up, and near one of these the Austrians left some planks, which enabled the French easily to restore the roadway. The great importance of proper arrangements for destroying bridges thoroughly and with certainty is thus forcibly illustrated by the operations of one day."

General instructions had been issued to the Austrian engineers in 1858 for mining the new bridges and viaducts, in readiness for blowing them up, as well as a special order to prepare a project for destroying the old bridge at San Martino. It was, therefore, proposed to bore cylindrical holes 8 inches in diameter horizontally across it, over two of the arches, according to a plan recommended in the instructions. This operation was, however, considered objectionable, because it might affect the stability of the arches, and explosion would not be certain to destroy the bridge, for the chamber of the mine having but little brickwork over it the explosion might take effect in an upward direction only; it was, therefore, subsequently proposed to sink two shafts, each 30 inches square, in the direction of the longitudinal axis of the central pier of the bridge which was to be destroyed. From the bottom of each shaft, galleries 30 inches high and 24 inches wide (the Austrian inch being very little longer), were to move in both directions, that is, four galleries in all. Each gallery was to terminate in a chamber to contain 62 pounds of gunpowder, and the centres of these were to be 9 feet from each other, the line of least resistance being equal to half the thickness of the pier, or 6 feet 4 inches.

The Committee of Engineers ordered to report on this plan made a third proposal. The shafts, galleries, and chambers were to be replaced by four cylindrical holes 7 inches in diameter, bored vertically, at the bottom of each of which there was to be a chamber 6 inches in diameter. Seventy pounds of 40° powder were required for each hole, or 280 pounds for each pier, this calculation having been based on the presumption that coefficient g should be taken at 0·255, the same as a compact rock.

The Committee were induced to make this proposal because they considered the making of shafts, galleries, and chambers too difficult, and likely to affect the stability of the bridge, an apprehension shared by all the architects who were consulted.

These subjects were thoroughly considered and reported on, and some variations were proposed. Suitable instruments, however, for boring the holes were wanting, and none could be procured at the moment, owing to the breaking out of the war. The chambers were blasted before with gunpowder, and the holes made much larger than necessary, and of very irregular shape. To meet this evil and produce the necessary width and smoothness, a pole had to be let down into each, and the interval between it and the masonry was filled up with bricks. The expenditure of powder used in this blasting was very much regretted subsequently, when another supply could not be procured, and it would then have been of essential service.

When orders arrived during the night of the 2nd of June to blow up the bridge, the chambers were filled with 86 pounds of powder in each, and partly tamped with sand; but it has not been ascertained whether the upper part was properly filled with clay and fragments of stone. On ignition, five only of the mines exploded, with a rather strong detonation, including the four in the first pier, but only one in the second.

After this unexpected failure the remaining charges were

at once examined, and being found intact were successfully exploded by means of electricity, and it was expected that the arches would then fall.

The second detonation equalled the first, but the arches did not fall. The masonry below them had been destroyed, and they sunk at the key stones two or three feet; the road across the bridge formed a concave line and was full of fissures; and the balustrades, formed of blocks of stone, had been precipitated into the river.

As there was no powder in reserve, two guns were ordered to batter the parts of the bridge most injured, in order to cause the arches to fall, but without success.

The account here shows in one part a want of the implements necessary for making the openings as determined upon, a deficiency at least in the requisite quantity of powder, doubts expressed as to the powder having a proper degree of strength, and a failure of the ignition of three charges out of eight; all of which difficulties, it is apprehended, might have been provided against by adopting a more simple off-hand mode of proceeding.

In destroying a bridge in a campaign, the system should vary with the time and means available; it is rare that that which is most correct theoretically can be adopted, and generally every disadvantage may be counteracted by employing plenty of powder. The object being the destruction, it is of far less consequence to have that in the extreme than to fail in obtaining what is required. (Extracts) *Vide* Papers of the Royal Engineers, pp. 116—120, vol. 10, 1861.

J. W.

ADDENDA.

The following interesting episode is added without reference having been made to the gallant officers referred to, the publisher claiming it as a subject belonging to our service, and more particularly as it is an acknowledged evidence of the distinguished services of the Corps of Royal Engineers of Great Britain and Ireland.

THE SERVICES OF THE ROYAL ENGINEERS IN THE CRIMEA.

SPEECH MADE IN THE HOUSE OF COMMONS BY

Captain L. Vernon.—I rise to call the attention of the House to the services of the Corps of Royal Engineers in the Crimea. I do so because there is a disposition abroad to depreciate the services of the British army in the Crimea. We think it high time that something should be done to counteract this tendency to detract, and we believe that my statement this evening will be a step in that direction. No detractor has ventured to question the courage and the conduct of the British soldiers of the general service. So far as the special corps are concerned, I have never heard any one bold enough to say the British artillery was second to any in the world, and my statement this evening will show that the British Engineers were equal, to say the very least, to any engineer that took the field.

The war that has just terminated, unlike any other modern war on record, narrowed itself into one mighty siege. The victory of the Alma was but the introduction to the siege of Sebastopol, and the battles of Balaklava, of Inkerman, and of the Tchernaya were but futile attempts on the part of the Russians to raise that siege. A fortress important rather for its uses than for its strength—a fortress so low in the scale of scientific defence that it was supposed, erroneously enough, to be open to a surprise, so moderately fortified that it was considered liable to the affront of a *coup de main*, became, under the pressure of circumstances and by the mere force of earthworks erected by the genius of Todtleben, one of the strongest places on record, and held at bay for eleven months the chivalrous valour and the military science of the world. This war, then, being a siege, it follows that the battle was fought by science. It was a war of engineers, and I rise in my place to claim for the British Engineers their full share in the achieving of that great result which has brought about the peace.

There were three great turning points on which the success of the war depended. First was the selection of a place of landing in the Crimea; secondly was the decision as to which front of Sebastopol should be attacked—for we were not in a condition to invest the whole, according to the real acceptation of the term; third, and most important, was the discovery of the key to the position of the front to be attacked.

I may at once avow and claim for the British Engineers the decision on all these three points, and I shall confine myself as much as possible to proving that this was the case. I must trust to some indulgence while I place historically before them these three questions in their relative positions. It will be seen at a glance that this question widens itself from a corps question into a national one. What I now say, by the aid of the press will be spread far and wide. What will, doubtless, by many be impugned, and it therefore behoves me to start on a proper base, and to go on adding fact to fact in order to be able to defy all contradiction.

In January, 1854, on account of the appearances in the East, Colonel Vicars, with three engineers, left England to place themselves under the orders of Admiral Dundas, who commanded in the East. At Gibraltar Colonel Vicars was taken ill, and the command devolved upon Captain Chapman, now Colonel Chapman, whose distinguished services I have had occasion before to bring under the notice of the House. These officers joined the fleet in the Bosphorus, and were despatched to reconnoitre the strong position of Maidos, near the Dardanelles.

Now, at this juncture the home authorities were without any precise information with regard to the East. In this dilemma Sir John Burgoyne, whose high position as Inspector-General of Fortifications might well have excused him from the arduous undertaking, volunteered his services, at this inclement season, to proceed to the East to make military observations of such forces as should be sent by the allied French and English armies in support of the Turks, in the event of a war with Russia which then appeared imminent. His services were accepted with eagerness. On his way through Paris the Emperor Napoleon associated with him Colonel Ardant, an officer of French Engineers. These two proceeded together to the Dardanelles, and inspected the position of Maidos, and afterwards of Boulahir, preferring which latter the officers of engineers were withdrawn from Maidos to reconnoitre Boulahir, which they did in that inclement season, the snow being then deep on the ground.

Sir John Burgoyne and Colonel Ardant then proceeded to Constantinople to reconnoitre the position of Bujukchekmedji, about twelve miles from Constantinople, a strong position, intended to be made the base of operations and to cover Constantinople. Colonel Ardant went forward to examine the position of Kara-su, where strong lines of defence were available, connecting the Sea of Marmora with the Black Sea. Sir John Burgoyne meantime went to Shumla, to confer with Omar Pasha, and he reconnoitred and reported upon Varna. Thence he returned to England, leaving Colonel Ardant at Gallipoli.

While Sir J. Burgoyne was at Constantinople, there was presented to him a project for the defence of that town by certain French officers attached to the embassy—these lines of defence were to pass from the Sea of Marmora to the Golden Horn, and from that to the Bosphorus, passing within a mile of the suburbs of Constantinople. The ground was ably taken up, but Sir John Burgoyne at once pointed out that it was faulty, because it passed close to an enormous population and a city liable to conflagration as was Constantinople; but the principal objection was, that it abandoned to the enemy the Bosphorus, which was our only means of communication with the Black Sea. This plan of defence, therefore, was abandoned in favour of that of Kara-su, which in every point resembled the lines of Lisbon, with a similar advantage of the stronghold of Bujukchekmedji.

War was now declared; the allied army was sent to Gallipoli, and took up the intrenched post of Boulahir; they then proceeded to Constantinople, leaving a small force to occupy Gallipoli. The Russians having made no impression on the Danube, notwithstanding their vast military resources, and the allied armies having advanced to Varna, in support of the Turks, the proceedings of Sir J. Burgoyne and of Colonel Ardant were criticised as being too cautious and unenterprising, by taking up a defensive position for Constantinople and the Dardanelles; but it must be remembered that at that time the war had not begun, and it could not have been supposed that the Russians, who, in so arrogant a manner, had forced on the war, should have been held entirely in check by the Turks; and it was therefore requisite that Constantinople should be protected, and the Dardanelles, without which there were no means of communicating with the Sea of Marmora, the Bosphorus, or the Black Sea, which latter was at that time in the possession of the Russian fleet; in a word, it would have been impossible to trust an allied army in that country if such a strong position as Gallipoli and its adjacents had not been found. Such was the opinion of the Emperor Napoleon, and, what is more to my purpose, such was the opinion of Sir John Burgoyne.

In August Sir John Burgoyne was sent out to command the engineers in the Crimea, and was placed upon the staff. In September the army embarked at Varna for the purpose of invading the Crimea.

And now I come to the first point I wish to prove—namely, the selection of the part of the Crimea in which the landing was to be effected. A council of war assembled on board the Caradoc. It was attended—on the part of the French, by General Canrobert, by Colonel Trochu, one of the French staff, and by General Bizot, the French engineer; on the part of the English by Lord Raglan, by Sir G. Brown, by Sir E. Lyons, and by Sir J. Burgoyne. The French held the opinion that the best place to land was at the mouth of the Katcha, and I believe that Sir G. Brown coincided with that opinion, but he said, "Before coming to a decision on this point, I think we ought to know the opinion of Sir J. Burgoyne, who has had more practical experience than any other officer present." On this Sir J. Burgoyne declared that the Katcha was not the proper place to land, that it was a difficult and defensible ground, and close to the resources and reserves of the Russians, and he pointed out, on the other hand, that the safest place to land was at the Old Fort.

Sir J. Burgoyne's representations were made known to

Marshal St. Arnaud, who at once grasped the idea and consented to the move. The landing, therefore, was safely effected at the Old Fort, and Eupatoria, in the rear, was seized and occupied. The abandoning of the idea of landing at the Katcha was very distasteful to some officers of the French staff, but when that place fell to our position it was seen that Sir John Burgoyne's estimate of the difficulty was right, and that an attempt to land there would have been followed by failure and disaster. I think I have proved now my first point, and that I have a right to claim the selection of the place for landing for the British Engineers.

I now come, sir, to my second point—that is, the selection of the side on which Sebastopol was to be attacked. After the battle of the Alma the troops advanced towards Sebastopol across the rivers Katcha and the Belbek. Now, the intention of the French, and for which they had prepared projects, was to attack Sebastopol on the north side. Sebastopol on the north side was situated on a promontory, and its defences were placed on rocky heights, having in front of them strong ground of a very defensible character, narrowed by the bay of Belbek on one side, the head of the harbour on the other, the promontory being dominated by a strong permanent work called the "Severnaia." Sir J. Burgoyne did not think that the north side of Sebastopol was the side to be attacked; he rather held to the opinion that it should be attacked on the south side, and he wrote a report to Lord Raglan, giving his reasons for holding that opinion, an extract from which report I will now, with the permission of the House, proceed to read:—

"The communications with the fleet, whence all resources were necessarily obtained, would be from the fine bays and harbours of Balaklava, Kamiesch, and Kazatch, instead of from an entirely open beach, which was alone available on the north. The fronts that were exposed to attack were extensive, and, though naturally of great strength, were not more so than that of the north, which was limited, and consequently admitted of defence after defence. The south side covered the docks, barracks, and all the great establishments of the place; whereas, if the north promontory were obtained, there was the harbour still intervening, which could not be crossed by any means; and the only resource would have been a bombardment, and not possession.

"In rear of the encamping ground to be occupied by the

allies in front of Sebastopol on the south side was a compact and most powerful position facing the country, and the communication to it from the harbours was direct and comparatively short, while on the north there was no favourable position on the land side; the ground to cover the camp and landing-place must have been of enormous extent, for that landing could not have been nearer than the Katcha, as the Belbek was commanded by the enemy's batteries, and the communication would have been much longer, and over two heights instead of one. The enemy, if attacked on the north, having but one front of the garrison, of moderate extent, to cover, could have greatly increased the outer field army for raising the siege. In thoroughly reconsidering every circumstance, it is impossible to conceive how the operation could possibly be sustained against the north side; nor how the army, were it to remain there, could avoid some frightful catastrophe."

This report was sent to Marshal St. Arnaud, and that officer, with his usual sagacity, accepted the idea, and consented to attacking Sebastopol on the south side. Then came the question, how was that to be done? If there be one axiom in war more cogent than another, it is that an army should never separate itself from its base; and if there is any other axiom equal to that in cogency, it is that a flank march should never be made in the presence of an enemy. And yet, at first sight, it would seem that the proposition of Sir J. Burgoyne embraced both these military errors; but it was not so in fact. He proposed to leave one base, but the base moved, so that he should fall upon it again; and the flank march to enable him to reach the south side of Sebastopol was in the rear of a flying and disorganised enemy, and it thrust the army between Menschikoff and Sebastopol. The movement was therefore undertaken, and the army sat down before Sebastopol, never to rise from it again till it left that place and its defences a shapeless ruin.

I think that I am entitled to say that I have proved my second point, and that I have a right to claim the selection of the side on which Sebastopol should be attacked for the British Engineers. The siege was now commenced with scanty military means. There were 300 or 400 sappers where there should have been as many thousands—for it should be remembered that behind the earthworks at Sebastopol was ranged the whole military power of Russia—

and where, if there had been as many thousands, it would have saved thousands of lives and millions of money.

There were 80 officers of engineers sent to the Crimea; of these 43 were killed, wounded, and put *hors de combat*—a wholesale slaughter with no parallel. Many of these officers passed in that inclement season, and under what the French call "fire of hell," 100 nights, making nearly a third of the whole time of the siege. Under that fire the executive officers, Chapman and Gordon, erected batteries of so substantial a character that they were not damaged by the fire of the enemy. The British artillery destroyed the fire of Todtleben, the Russian artillery swept from the face of the earth the French batteries, but no missile hurled against the English batteries stopped for one single moment their steady, sure, and onward course. From the first reconnaissance of Sebastopol, Sir J. Burgoyne perceived that the Malakhoff was the key to the position of the front attack, and he so represented it to Lord Raglan.

After the battle of Inkerman he again impressed on the authorities that the Malakhoff was the place to be attacked. Upon the arrival of General Niel, the French aide-de-camp of engineers to the Emperor, a council of war of the allied engineers was held: at that council of war Sir J. Burgoyne again represented that the Malakhoff was the key to the position, and that it should be attacked. After the council of war had been held, wishing to place on record his opinion, he reduced it to writing, and, through Lord Raglan, sent it to the French engineer, General Niel. The following day General Niel called a council of French engineers to take under consideration Sir J. Burgoyne's memoir—they prepared a *procès verbal* of what there took place, and sent a copy of it to Lord Raglan for Sir J. Burgoyne's information. The first paragraph of that *procès verbal* stated that the Malakhoff should be attacked in compliance with the opinion of Sir J. Burgoyne. The words used were these:—

"Il résulte des dispositions adoptées en conseil, et suivant le vœu exprimé par le Lieutenant-Général Sir J. Burgoyne, que des travaux d'approche devront être exécutés devant la tour Malakhoff, afin de pouvoir attaquer, par ce point dominant, le faubourg de Karabelnaia, en même temps qu'on donnera l'assaut à la partie ouest de la ville."

I think, therefore, I have made out my third point, and that I am justified in claiming the discovery of the key to the front attacked for the British Engineers. Now that I have established the claim of the British Engineers to the merit of deciding on the three turning points of this war—they forming a part, and an important one, of the British army—what becomes of the case of those who would seek to depreciate the services of the British army in the Crimea?

APPENDIX.

Mechanical Rock Boring and Cutting.

Since the date of Sir John F. Burgoyne's work on blasting, improvements of the highest importance and interest have been made and practically brought into operation for perforating the holes (jumper holes) for blasting in rock, or for actually *hewing* or cutting out into blocks the softer rocks; such operations being indispensable preliminaries to blasting, although, as in the case of coal-cutting by machinery, they need not be followed by the use of gunpowder.

We can only incidentally notice these improvements, and refer the reader to works in which an account of them, or some of them, may be found; for to describe them in detail even imperfectly would be to greatly increase the size of this volume.

Although many anterior attempts had been made, and with some success, to produce a mechanical rock perforator, it was the necessity for piercing a railway tunnel through the Alps at Mont Cenis that, between the years 1855 and 1860, gave the decisive impetus to invention for the purpose.

The earliest successful machine was produced by the late Mr. Thomas Bartlett, C.E., an English civil engineer, employed as chief agent under Messrs. Brassey and Co., on the Victor Emmanuel Railway. It was exhibited in action, with steam as the motor, and afterwards with compressed air as such, before Count Cavour and other members of the Italian government, perforating rock of various sorts and hardness, in the neighbourhood of Genoa. Although the credit as well as the profits accruing from this great advance have been seized by others, to the exclusion of the originator, the machines at this time employed at Mont Cenis Tunnel are in principle and almost in every point of construction identical with those of Mr. Bartlett. The machines at this tunnel perforate several jumper holes simultaneously, and at more than three times the rate that they could be executed by hand labour. Various other perforating machines, more or less analogous to these, have been since produced, some of which are in successful use in mining operations in Sweden and elsewhere.

The best accounts of the machines in all their details will be found in the following works:—

1. " Relazione della direzione Tecnica, alla direzione generale delle strade ferrate dello stato; sullo Traforo delle Alpi tra Bardonnèche e Modane." 4to. Torino, 1863.

2. "Une Visite a la percée du Mont Cenis." Par Paul Eymard, &c. 8vo. Lyon, 1863.

3. "Notice Historique sur la percée du Mont Cenis." Par J. Bonjean. 8vo. Pouchet, a Chambéry, 1863.

4. "Notice Historique et Critique sur les Machines de Compression d'Air du Mont Cenis," &c. Par le Marquis Anatole de Caligny. 4to. Turin, 1860.

5. "Theorie du compresseur a Colonne d'Eau, de MM. Sommeiller, &c.; fonctionant au percement des Alpes." Par M. de Saint-Robert. Mémoire, Annales des Mines. 6ieme série, tom. iii., 1863.

6. "Rapport du Bureau Central, &c., sur le Projet de loi, pour la percée, &c., de Senato del Regno." Sessione Parlamentare, del 1857. No. 68, bis. (An Italian Parliamentary Paper. 8vo. Turin.)

7. "Application de la Théorie Mécanique de la Chaleur, au Compresseur Hydraulique du Tunnel des Alpes." Par M. A. Cazin. Annales des Mines.

To these we have to add the following in English:—

1. "Account of the Works of the Mont Cenis Tunnel," by M. F. Sopwith; which will soon appear in the Minutes of Proceedings of the Institution of Civil Engineers, before which it was read, Session of 1864—65.

2. "The Practical Mechanics' Journal," in which several articles in the years 1863—64—65 will be found on rock perforating and coal-cutting machines.

As regards tools for rock especially hard rock perforation, it may be stated as a fundamental principle that no scraping or cutting tool that proposes to act by uniformly continued slow movement can ever have even a chance of success as a rock perforator. For this, in one form or another, *percussion* is indispensable, while rocks continue to consist in part of silex, and our available tools of hardened cast steel. Strike the rock, and the rigid but brittle material is shattered more or less, while the less rigid but tougher cast steel suffers but little; attempt to scrape away or cut the rock, and the softer material, the steel, is rapidly ground away, and almost no impression made upon the harder one, the rock. This obvious fundamental principle seems to have been unknown or disregarded in all earlier attempts.

The Mont Cenis Tunnel will be 12,220 metres, or 7·5932 English miles, in length.

The French end entrance is 3,945 ft. above the Mediterranean, and the Italian entrance 4,379 ft. above the same; while the summit above of the Alpine ridge is 9,525 ft. above the sea level.

The tunnel will be 26 ft. in width at the broadest, and 25 ft. 3 in. at the level of the rails, with a height of 24 ft. 7 in., the transverse section being something of the shape of a compressed parabola, though formed of circular segments in its curved perimeter.

The rocks through which it is being driven, are of the following sorts, taking our departure from the Fourneau, or Modane end, at the north, viz.:—

1. Micaceous and talcose sandstones, slaty and twisted in structure.
2. Quartzites, in which it was expected to have encountered a thick bed of very hard quartz, but which will probably now not be met with, as the expected point of its existence, as indicated from above, has been already passed.
3. Formations of gypseous rocks, with much anhydrite.
4. Limestones of fine grain, slaty structure, and of great hardness, in which occur some seams and beds that look at first sight almost like anthracite, but which are probably only metamorphosed black-shining carbonaceous shales.
5. Various argillaceous and calcareous schists, with siliceous veins, &c.

The power employed at Mont Cenis is that of water. As it falls from the opposite sides of the mountain, it is intercepted at a sufficient height to afford at the Bardonnèche end a head of 85 feet; a less one only could be obtained at the Modane end. Two sorts of machines are in operation, by which air is compressed to a tension of five atmospheres by means of this water power. At the Bardonnèche end these are of the sort called "*à coups de bélier.*" They are in fact gigantic hydraulic rams, identical in principle with that of Montgolfier for raising water, but having this difference, that here the entire work—accumulated in a given volume of water caused to descend through a given fall at each stroke—is entirely expended in compressing a certain volume of air from atmospheric pressure to that state.

Each machine is a U-shaped syphon pipe, furnished with certain valves, so arranged that the descending column of water in one leg, rising into the other, compresses the air pent between it and the valves, which, near the top, discharge into a large air-vessel. When the one oscillation of the descending column of water has done this work, it is discharged by another valve at the lower part of the syphon, and its place filled with air ready for another compression by the next stroke. There are twelve or more of these *béliers*, all similar, working into a common reservoir, which communicates by a pipe $7\frac{1}{2}$ inches diameter with the remotest interior end of the tunnel working or forehead. The main pipe of each *bélier* is 2 feet diameter, with, as has been said, 85 feet of head, and communicating with an air-vessel of 600 cubic feet of capacity. Such machines alone were employed at first. Subsequently at both ends of the tunnel, air-compressing pumps, driven by water-wheels, have been superadded. At the Modane end 85 feet of head, which is required to compress the air to 5 atmospheres, not being afforded by nature, water-wheels were established by the engineers, MM. Sommeiller et Grandis, to pump water up to that elevation, whence it was let down again in acting upon the *béliers*—a most extraordinary example of mistaken dynamics, and one that we should not have been prepared to find practised by men of their undoubted ability. More recently, however, this has been altogether abandoned at this end, and the water-wheels applied directly to compress the air by air-pumps.

These pumps have two horizontal cylinders, each 21½ inches diameter, with a 5-foot stroke, and making 12 strokes per minute. The pistons do not act upon the air directly, but by the intervention of an oscillating short column of water held above them, whereby leakage of the air past the piston is avoided in great part. The air thus compressed and introduced to the forehead, is then employed (exactly as in the coal-cutting machinery of Frith and Donisthorpe, described in the "Practical Mechanics' Journal," for 1864), to actuate small perforating machines, travelling upon a framework on wheels and railway, laid in the tunnel, beneath which is the air-pipe. Each machine is much of the nature of a small high-pressure horizontal steam engine—the prolonged piston-rod of which is the steel drill or jumper, by the repercussion of which the rock is perforated for the admission of the charges of powder by which the forehead is successively blown away.

Each machine is so fixed to the frame that the direction of the blows of the jumper can be changed and fixed at will, and it embraces movements by which the jumper is twisted slowly on its own axis, and advanced as the hole deepens. Each machine makes 250 strokes per minute, with a stroke variable from 2 inches up to 7 inches. The striking *plus* pressure on the piston, or that over and above the back pressure, or air-spring which brings back the piston at the end of the stroke, is 216 lbs.; so that, at full stroke, each machine is equivalent to 2·4 horse-power in action.

From the 7½-inch main, flexible pipes of vulcanised India rubber, wrapped with wire (to enable it to withstand the pressure), and of 2 inches internal diameter, supply the air to each machine.

There are eleven of these *percuteurs* or perforators at one end, and nine of them at the other, usually at work.

A large close vessel of water accompanies the carrying-frame at each end, and is in communication with the air-main, so that a stream of water under a pressure of five atmospheres is constantly squirted into each jumper-hole while in progress, and washes out the *débris* away from the cutting-edge.

Each machine weighs 6 cwt. They are so frequently shaken to pieces, or damaged more or less by the recoil of the rough jarring blows to which they are constantly exposed, that it is necessary to have four machines always under repair for one in action.

About eighty jumper-holes are bored *parsemé* over the face of the forehead by changing the position of the several machines to each shift. When these are bored, each from two-thirds to a metre deep, the whole frame with the machines is drawn back on the rails about 100 yards, the holes are all charged by cartridges, and fired nearly simultaneously. A pair of temporary folding-doors, in advance of the place of the machines, are closed, to prevent injury by chance projectiles. The *déblais* is removed in small waggons, on side rails, as soon as possible after the explosion, and then the frame and machines are run up again, to begin work anew.

A much larger jumper-hole is made in the centre of the face, and it is thought this central heavy charge is of great value in removing a greater depth of rock. There can be no doubt of this within certain not very large limits. We believe the very best effect would be obtained by firing all the charges *absolutely* simultaneously by galvanism, for which instruments are provided of great simplicity and certainty.*

As regards the comparative cost of this method and of handwork, Mr. Sopwith, in his paper read before the Institution of Civil Engineers, concludes that it is two and a half times as costly as the same result obtained by hand labour, but that the rate of progress is *three times as fast*. We must say we were unable to agree with the former part of Mr. Sopwith's conclusion, and believe that his estimate greatly overrates the cost. To justly estimate the comparative cost, the time of final completion is an indispensable element, in the enormous sums saved in interest on capital and in salaries, &c., &c., on the one hand, and by hastening of the period for returns by traffic begun, upon the other.

From 1858, when the works at Mont Cenis were nominally begun, to the end of the year 1862, the average rate of progress of the tunnel was 3,686 feet per day. This soon after reached and exceeded 4 feet per day. Still more recently, however (March, 1865), a comparatively soft rock has been entered at the Modane end, and whereas prior to this the best rate of progress anticipated was at the average of 800 metres per annum, or of 400 for either end, the experience at the Modane end has since been that this soft rock is penetrated at the rate of 250 metres per month, or at that of 3,000 metres per annum. If this should be confirmed and continued, the aperture right through may be looked for within the next three years, or three and a half. In such an event, what becomes of any estimate of comparative cost which leaves out of view the *saving* of perhaps *six years'* interest on capital, &c., and substitutes on the other side probably five and a half years' profits of traffic?

The cost so far (*i.e.* prior to the soft rock being found) would appear, according to Mr. Sopwith, to be about £210 per metre forward on the French end. The total estimated cost of the work is £2,600,000, as fixed by the engineers of the lines at the French and Italian sides, and by the convention as to payment, entered into by these respective governments.

The jumper-holes are charged with powder in waterproof cartridges, and fired either by fuse or galvanically. For some advantageous arrangements invented and employed for the galvanic ignition of blasting charges in jumper-holes, by Mr. R. Mallet, C.E., reference may be made to his paper on the subject, printed in the Transactions of the Institution of Civil Engineers of Ireland.

On the subject of rock-*hewing* machinery, *i.e.* machinery for cutting out thin dividing channels, in coal particularly, but equally applicable to soft rock, such as our Portland or Bath stone, or the Caen stone, the reader will find a good deal of information in the

---

* Such instruments, the invention of Mr. Becker, are made for sale by Messrs. Elliott, St. Martin's Lane, London.

"Practical Mechanics' Journal" for 1863—64, in the "Transactions of the Northern Mining Institute" for March, 1865, and in the printed specifications of the Patent Office. Not less than forty or fifty different patents for such machinery have been already taken out in Great Britain only, and many abroad. Amongst these various machines, Firth and Donnisthorpe's, which works with compressed air, is practically one of the most successful. It has been tried in Yorkshire and also in Durham coal-seams with perfect success. The hydraulic coal-cutting machine patented by Messrs. Carrett and Marshall, of Leeds, engineers, has also been worked with success in the Kappax Colliery, near Leeds. It isolates the blocks of coal ready for removal at about one half the cost of manual labour, and saves more than 18 per cent. of the coal at present *wasted totally* by the process of hand hewing.

This machine works upon a principle very analogous to a slotting machine tool, the cutting instruments being driven forth and retracted by water pressure, and moving at a moderate velocity. This, like the former machine, will work well in coal, or *soft, unsiliceous* rock, but upon the fundamental principle already adverted to, this last machine cannot be employed in any hard or gritty rock, or in one of very uneven or crystalline texture.

It would be foreign to the direct subject of this volume, however, to go more at length into this latter class of machines, as being but indirectly connected with the subject of blasting.

In another and wholly different direction a good deal of additional light has been thrown, since the period of Sir F. Burgoyne's writing, upon the advantages (under certain circumstances of rock) of substituting for jumper operations the use of great *mines* for quarrying purposes.

Thus, at Holyhead Quarries, where the rock is quartzite, mines occasionally of four or five headings, and charged with six or seven tons of gunpowder, have been fired, and mines of one or two tons are a thing of daily routine for years past.

The economy in labour thus effected in procuring the rock, and the advantage in obtaining very large blocks (fifteen tons being aimed at), are very great. Some account of these operations has been given, we believe, by Mr. Cousins, C.E., the intelligent agent of Messrs. J. and C. Rigby, the contractors for the Asylum Harbour, at Holyhead, for the formation of which the quarries are wrought. Since the appearance of this work, also, the whole subject of gun-cotton, as an explosive agent, has greatly advanced, and there appears good ground for concluding that ultimately, and ere very long, it may be found a far more econonomical and safe, as it is undoubtedly a far more powerful explosive agent for blasting, or for mines in rock, than the ancient agent, gunpowder.

Upon this part of the subject the "Transactions of the British Association" for 1863, 1864, and 1865, the Report of General Lenk, of the Austrian service, and various military and artillery journals, British and foreign, during the last three years, may be consulted. R. M.

INDEX.

Aberdeen, blasting in the granite quarries of, 78.
Air pumps used in tunnelling Mont Cenis described, 130.
Angers, slate quarries at, 83.
Apparatus, Vignoles boring, 5.
Arrangement of charges for blasting, 7.
Arrows, tamping, 53.
Augers, use of for boring, 4.
Baking, slate hardened by, 83 n.
Bangor, slate quarries at, described, 82.
Bartlett's rock perforating machine, 127; works containing accounts of, 127.
Batteries, firing mines by voltaic, 84; firing trains by voltaic, 34.
Bickford's patent fuse described, 30 n.; use of, 30.
Bickford's sump fuse, 32; how used, 32.
Blasting: advantages derived from joints in the rock, 14; boring for, 2; results of boring in the granite quarries at Dalkey, 3; boring out tamping, 4; charges of powder for, 6, 24; arrangement of charges, 7; proper method of arranging the charges in various cases, 9; usual charge, 9; firing, 31; firing in shafts, 35; firing by voltaic battery, 34.
Blasting in deep water, 32; in the granite quarries of Aberdeen and Peterhead, 78; at the Holyhead quarries, 131; in the Jumna, and at Delhi, 76.
Blasting limestone of the Antrim coast, 78; limestone at Plymouth, 77, 79.
Blasting, loading for, described, 26.
Blasting at Mont Cenis, 131.
Blasting needle, 2.
Blasting powder, experiments on the power of, 18; experiments on the strength of Government and merchants' powder, 20; results, 21
Blasting, powder best for, 22 quality of powder for, 17; space occupied by a given quantity of powder, 16; quarry shields, 36; selection of holes for, 5; system of, at Gibraltar, 25; tamping, 26, 37; ordinary process of tamping, 27; evils of the ordinary process of tamping, 27; improvements on the ordinary process of tamping 29; tamping plugs, 50; tools for, 2; use of nitric acid to form a cavity at the end of a bore, 74 and 16 n.; value of leaving a hollow space above the charge in, 23; Vignoles boring apparatus, 5
Blasting train, described, 29.
Boring apparatus, Vignoles, 5.
Boring for blasting, 2; churn jumpers, advantage of, 4; danger of reboring a charged hole, 4; holes, selection of, for boring, 5; improved methods of, 127; method of enlarging the inner end of bores with nitric acid, 16 n. and p. 74; boring out tamping, 4; use of augers for boring out tamping, 4; results of boring in the granite quarries at Dalkey, 3; waste of steel and iron at ditto, 3; weight of hammers used at ditto, 3.
Bridge at Duenas, destruction of, described, 111.
Bridges, attempted destruction of Hungarian, 115; attempted destruction of, over the Ticino, 117; method of mining, 109; Spanish, described, 109; system followed in the destruction of, in Spain, 109; wooden, system of destroying, 114.
Brick, broken, value as tamping, 46; experiments on, 47; conclusions from ditto, 49,
Building, stones used in, 88; division of, 89; granite, 89; limestones, 101; magnesian limestones, 102; results of Professor Daniell's experiments on, 105; sandstones, 97 serpentine and porphyry, 95; slate. 96.
"Bulling" defined 79.

INDEX.

Carrett and Marshall's coal-cutting machine, 131.
Charges, arrangements of, for blasting, 7; proper method of arranging the, in various cases, 9; usual, 5.
Chisel quarrying, 2.
Churn jumpers, 3.
Clay tamping, boring out, 4; table showing the value of clay and sand for, 37.
Cliff, removal of the Round Down, Dover, 84.
Coal cutting machines, 131.
Commercial-road stoneway, experiments made on the, to test the wear of granite, 93.
Cones, use of, in tamping, 51.
Cornwall granite, described, 90.
Cost of tunnelling, 72.
Crimea, services of the Royal Engineers in the, 120.
Dalkey, results of boring in the granite quarries at, 3.
Danger of reboring a charged hole, 4.
Daniell's experiments on the stones used in building, 105.
Delhi, blasting at, described, 76.
Destruction of bridges in Spain, described, 109.
Double tunnels, 63; comparison of the advantages of single and double, 64.
Dover, removal of the Round Down Cliff at, 84.
Drainage of tunnels, 68.
Drenodrohur, tunnel at, 72; cost of ditto, 73.
Duenas, destruction of the bridge at, described, 111.
Eisleben, slate quarries at, 83.
Electricity, firing mines by the aid of voltaic, 84.
Engineer department, powder used by the, for blasting, 22.
Engineers, services of the Royal, in the Crimea, 120.
Eprouvette mortar, described, 19.
Eprouvette gun, described, 19.
Experiments on the power of blasting powder, 18; on the strength of Government and merchants' blasting powder, 18; results, 21.
Felspar, composition of, 89.
Firing mines by voltaic electricity, 84.
Firing trains by voltaic battery, 34.

Floors of tunnels, 65.
France, slate quarries in, 83.
French method of ventilating shafts, &c., 71.
Fuse, Bickford's patent, 30; use of, described, 30; sump, described, 32.
Germany, slate quarries in, 83.
Gibraltar, system of blasting at, 25.
Gneiss, 96.
Granite described, 89; celebrated works constructed of, 90; constituents of, 89; Cornwall, described, 90: Ferm, described, 90; porphyry, 95; quarries of Aberdeen and Peterhead, blasting in the, 78; resistance to crushing of the various kinds of, in use, 93, 94; results of boring in, 3; Scotch, described, 90; serpentine, 95; table of the results of the wear of different granites, 92; Vale stone, 92; water, absorption of, by Irish granite, 95; wear of, 93; where found, 89.
Gun-cotton as a blasting agent, 131.
Gun, éprouvette, 19.
Gunpowder, see *Powder, blasting*.
Guernsey granite, described, 90.
Herm granite, described, 90.
Holes, boring, for blasting, 2; danger of reboring charged, 4; selection of, for boring, 5; table showing the space occupied by a given quantity of powder in round, 16; use of nitric acid to enlarge the inner end of, 16 n. and 74.
Holyhead quarries, blasting at the, 131.
Hornblende, composition of, 89.
Hornstone, 96.
Hungary, attempted destruction of bridges in, described, 115.
Ilmenau, slate quarries at, 83.
Introductory observations, 1.
Irish granite, absorption of water by, 95.
Joints, advantage derived from, in rocks, 14.
Jumna, blasting in the, 76.
Jumpers, 2; churn, 3; advantage of churn, 3.
Kingstown Harbour, account of the use of the patent fuse at, 30.
Lavagna, slate quarry at, 83.
Liebhaber's (Baron), patent for enlarging the inner end of bores for blasting, 75.

INDEX. 135

Limestone, blasting the, of the coast of Antrim, 78; blasting, at Plymouth, 77, 79; cost of blasting, 80.
Limestones and oolites, 101; analyses of, 101; buildings constructed of, 102; composition of, 102; weights of, 101.
Limestones, magnesian, analyses of, 102; buildings constructed of, 103; composition of, 103; durability of, 103; weight of, 102.
Line of least resistance defined, 5.
Liverpool, cost of the railway tunnel at, 73.
Loading, process of, 26.
Materials most suitable for tamping, 59.
Mines, firing, by voltaic electricity, 84.
Mines, Mallet's method of firing galvanically, 131.
Mont Cenis Tunnel, 128; blasting at, described, 131; machinery, &c., employed, 129; air pumps, 130; strata to be passed through, 129.
Mortar, éprouvette, 19.
Needle, blasting, 2
Nelson's memoranda on the quarrying of Plymouth limestone, 79.
Nitric acid, use of, for forming a cavity at the end of a bore, 16 n, 74.
Observations, introductory, 1.
Oolites, see *Limestones.*
Packed sand, value of, for tamping, 44.
Patents for rock-hewing machinery, 131.
Percussion, necessity of tools for rock perforation acting by, 128.
Perforators employed at the Mont Cenis Tunnel described, 130.
Pesth, attempted destruction of the bridge at, 115.
Peterhead, blasting in the granite quarries of, 78.
" Plug and Feather," cutting with, 79.
Plugs, tamping, described, 50; results of experiments with, 55.
Plymouth, blasting limestone at, 77, 79; cost, 80.
Porphyry, 95.
Potstone, 95.
Powder, blasting: best for blasting, 22; charges, 5, 24; experiments on the power of, 18; experiments on the strength of Government and merchants', 20; results of ditto, 21; mixture of quicklime with, 24; quality of, 17; sawdust, its value in, 22; space occupied by any given quantity of, 16; used by the Royal Engineer department, 22; usual charge, 5.
Quality of powder used for blasting, 17.
Quarries, blasting in the, of Aberdeen and Peterhead, 78.
Quartz, composition of, 89.
Quartz rock, 96.
Quarry, rubbish, use of, as tamping, 46.
Quarry shields, 36.
Quarrying slate, 81; tools for, 2.
Quicklime, mixture of, with blasting powder, 24.
Resistance, line of least, defined, 5.
Rimogne, slate quarries at, 83.
Rock, best positions to operate on, 9
Rock, advantages of joints in the, 14.
Rock-hewing machinery. 131.
Rock perforating machinery, Bartlett's, 127.
Rock perforation, necessity of tools for, acting by percussion, 128.
Roofs of tunnels, forms of, 63.
Round Down Cliff described, 84; removal of, by blasting, described, 85; cartridges, 87; cases for the charges, 87; chambers for depositing the charges, 85; charges, 85, conducting wires, 86; firing the mines, 87; mass removed, 88; priming wires, 87; tamping, 87; voltaic batteries described, 86.
Sand, use of, for tamping, 38; experiments on tamping with packed, 44; conclusions from ditto, 45; table showing the comparative value of clay and sand for tamping, 40.
Sandstones, 97; beds of, 97; chemical analyses of, 97; durability of, 99; resistance of, to pressure, 99; selection of, for building purposes, 98; weight of, 98.
San Martino, attempted destruction of the bridge at, 117.
Sawdust, value of mixing, with blasting powder, 22.
Scotch granite described, 90.
Serpentine described, 95; where found, 95.

Shafts, sinking, 66; disadvantages of, 66; firing trains in, 35.
Shells, experiments on the strength of Government and merchants' blasting powder, by bursting, 20.
Shields, quarry, 36.
Slate, constituents of, 96; geological division of, 96; hardened by baking, 83 n.; Irish quarries, 97; quarries in England, 97; Scotch quarries, 97.
Slate quarrying, 81; at Bangor, 82; in France, 83; geological formation, 82; greenstone dykes, removal of, 83; in Italy, 83; process explained, 82; in Switzerland, 83.
Spanish bridges described, 109.
Stone tamping, boring out, 4.
Stone, broken, as tamping, 46; experiments on, 47; conclusions from ditto, 35.
Stones used in building, 88; division of, 89; granite, 89; limestones, 101; magnesian limestones, 102; results of Professor Daniell's experiments on, 105; sandstones, 97; serpentine and porphyry, 95; slate, 96.
Strata, effects of, in tunnelling, 67.
Sump fuse described, 32.
Switzerland, slate quarries in, 83.
Table of the power of blasting powder, 18; of experiments on the comparative strength of Government and merchants' blasting powder, 20; showing the space occupied by any given quantity of powder in round holes, 16; of the wear of different granites, 92; showing the resistance to crushing of the various kinds of granite in use, 93; showing the resistance to pressure of sandstones, 100; showing the value of sand and clay for tamping, 40; of experiments showing the value of packed sand for tamping, 44; showing the results of experiments on tamping with broken brick and stone, 47; showing the results of experiments with tamping plugs of different forms, 55.
Tamping described, 26, 37; boring out, 4; broken brick, 46; broken stone, 46; materials for, 37; observations on the materials most suitable for, 59; plugs, 50; results of experiments with plugs, 55; use of sand for, 38; experiments on the relative value of sand and clay for, 40; experiments on the value of packed sand for, 44; conclusions from ditto, 45.
Ticino, attempted destruction of the bridge over the, near Magenta, described, 117.
Tools for blasting, 2; for quarrying, 2.
Train for firing charges, 29
Trains, firing, 34.
Tremenhere, Lieut., blasting operations in the Jumna and at Delhi, 76.
Tunnelling: at Mont Cenis, 128; cost of, 72; cost of the road tunnel of Drenodrohur, 72; cost of the railway tunnel at Liverpool, 73; difficulties of, 60; dimensions of tunnels, 62; double tunnels, 63; comparison of the advantages of single and double tunnels, 64; drainage of tunnels, 68; effects of strata, 67; floors of tunnels, 65; form of roof of tunnels, 63; incline of tunnels, 68 n.; process of, 61; shafts, sinking, 66; arrangements for ditto, 66; disadvantages of ditto, 66; ventilation of tunnels, 69; apparatus for the ventilation of tunnels, 71.
Vale stone granite, described, 92.
Ventilation of tunnels, 69; apparatus for the, 71.
Vernon, Captain, on the services of the Royal Engineers in the Crimea, 120.
Vialet's method of hardening slate by baking, 83 n.
Vignole's boring apparatus, 5.
Voltaic battery, firing trains with the, 34; firing mines by, 84.
Water, absorption of, by Irish granite 95; blasting in deep, 32.
Whetstone slate, 96.
Wooden bridges, system of destroying, 114.

WEALE'S SERIES
OF
SCIENTIFIC AND TECHNICAL WORKS.

"It is not too much to say that no books have ever proved more popular with or more useful to young engineers and others than the excellent treatises comprised in WEALE'S SERIES."—*Engineer.*

A New Classified List.

	PAGE		PAGE
CIVIL ENGINEERING AND SURVEYING	2	ARCHITECTURE AND BUILDING	6
MINING AND METALLURGY	3	INDUSTRIAL AND USEFUL ARTS	9
MECHANICAL ENGINEERING	4	AGRICULTURE, GARDENING, ETC.	10
NAVIGATION, SHIPBUILDING, ETC.	5	MATHEMATICS, ARITHMETIC, ETC.	12
BOOKS OF REFERENCE AND MISCELLANEOUS VOLUMES . . 14			

CROSBY LOCKWOOD AND SON,
7, STATIONERS' HALL COURT, LONDON, E.C.
1897.

CIVIL ENGINEERING & SURVEYING.

Civil Engineering.
By HENRY LAW, M. Inst. C.E. Including a Treatise on HYDRAULIC ENGINEERING by G. R. BURNELL, M.I.C.E. Seventh Edition, revised, with LARGE ADDITIONS by D. K. CLARK, M.I.C.E. . . . **6/6**

Pioneer Engineering:
A Treatise on the Engineering Operations connected with the Settlement of Waste Lands in New Countries. By EDWARD DOBSON, A.I.C.E. With numerous Plates. Second Edition **4/6**

Iron Bridges of Moderate Span:
Their Construction and Erection. By HAMILTON W. PENDRED. With 40 Illustrations **2/0**

Iron (Application of) to the Construction of Bridges, Roofs, and other Works.
By FRANCIS CAMPIN, C.E. Fourth Edition . . . **2/6**

Constructional Iron and Steel Work,
as applied to Public, Private, and Domestic Buildings. By FRANCIS CAMPIN, C.E. **3/6**

Tubular and other Iron Girder Bridges.
Describing the Britannia and Conway Tubular Bridges. By G. DRYSDALE DEMPSEY, C.E. Fourth Edition **2/0**

Materials and Construction:
A Theoretical and Practical Treatise on the Strains, Designing, and Erection of Works of Construction. By FRANCIS CAMPIN, C.E. **3/0**

Sanitary Work in the Smaller Towns and in Villages.
By CHARLES SLAGG, Assoc. M. Inst. C.E. Second Edition . **3/0**

Roads and Streets (The Construction of).
In Two Parts: I. THE ART OF CONSTRUCTING COMMON ROADS, by H. LAW, C.E., Revised by D. K. CLARK, C.E.; II. RECENT PRACTICE: Including Pavements of Wood, Asphalte, etc. By D. K. CLARK, C.E. **4/6**

Gas Works (The Construction of),
And the Manufacture and Distribution of Coal Gas. By S. HUGHES, C.E. Re-written by WILLIAM RICHARDS, C.E. Eighth Edition . **5/6**

Water Works
For the Supply of Cities and Towns. With a Description of the Principal Geological Formations of England as influencing Supplies of Water. By SAMUEL HUGHES, F.G.S., C.E. Enlarged Edition . . . **4/0**

The Power of Water,
As applied to drive Flour Mills, and to give motion to Turbines and other Hydrostatic Engines. By JOSEPH GLYNN, F.R.S. New Edition . **2/0**

Wells and Well-Sinking.
By JOHN GEO. SWINDELL, A.R.I.B.A., and G. R. BURNELL, C.E. Revised Edition. With a New Appendix on the Qualities of Water. Illustrated **2/0**

The Drainage of Lands, Towns, and Buildings.
By G. D. DEMPSEY, C.E. Revised, with large Additions on Recent Practice, by D. K. CLARK, M.I.C.E. Second Edition, corrected . **4/6**

Embanking Lands from the Sea.
With Particulars of actual Embankments, &c. By JOHN WIGGINS . **2/0**

The Blasting and Quarrying of Stone,
For Building and other Purposes. With Remarks on the Blowing up of Bridges. By Gen. Sir J. BURGOYNE, K.C.B. **1/6**

Foundations and Concrete Works.
With Practical Remarks on Footings, Planking, Sand, Concrete, Béton, Pile-driving, Caissons, and Cofferdams. By E. DOBSON, M.R.I.B.A. Seventh Edition **1/6**

Pneumatics,
   Including Acoustics and the Phenomena of Wind Currents, for the Use of Beginners. By CHARLES TOMLINSON, F.R.S. Fourth Edition . **1/6**

Land and Engineering Surveying.
   For Students and Practical Use. By T. BAKER, C.E. Fifteenth Edition, revised and corrected by J. R. YOUNG, formerly Professor of Mathematics, Belfast College. Illustrated with Plates and Diagrams . . . **2/0**

Mensuration and Measuring.
   For Students and Practical Use. With the Mensuration and Levelling of Land for the purposes of Modern Engineering. By T. BAKER, C.E. New Edition by E. NUGENT, C.E. **1/6**

MINING AND METALLURGY.

Mineralogy,
   Rudiments of. By A. RAMSAY, F.G.S. Third Edition, revised and enlarged. Woodcuts and Plates **3/6**

Coal and Coal Mining,
   A Rudimentary Treatise on. By the late Sir WARINGTON W. SMYTH, F.R.S. Seventh Edition, revised and enlarged **3/6**

Metallurgy of Iron.
   Containing Methods of Assay, Analyses of Iron Ores, Processes of Manufacture of Iron and Steel, &c. By H. BAUERMAN, F.G.S. With numerous Illustrations. Sixth Edition, revised and enlarged . . **5/0**

The Mineral Surveyor and Valuer's Complete Guide.
   By W. LINTERN. Third Edition, with an Appendix on Magnetic and Angular Surveying **3/6**

Slate and Slate Quarrying:
   Scientific, Practical, and Commercial. By D. C. DAVIES, F.G.S. With numerous Illustrations and Folding Plates. Third Edition . . **3/0**

A First Book of Mining and Quarrying,
   with the Sciences connected therewith, for Primary Schools and Self Instruction. By J. H. COLLINS, F.G.S. Second Edition . . . **1/6**

Subterraneous Surveying,
   with and without the Magnetic Needle. By T. FENWICK and T. BAKER, C.E. Illustrated **2/6**

Mining Tools.
   Manual of. By WILLIAM MORGANS, Lecturer on Practical Mining at the Bristol School of Mines **2/6**

Mining Tools, Atlas
   of Engravings to Illustrate the above, containing 235 Illustrations of Mining Tools, drawn to Scale. 4to. **4/6**

Physical Geology,
   Partly based on Major-General PORTLOCK's "Rudiments of Geology." By RALPH TATE, A.L.S., &c. Woodcuts **2/0**

Historical Geology,
   Partly based on Major-General PORTLOCK's "Rudiments." By RALPH TATE, A.L.S., &c. Woodcuts **2/6**

Geology, Physical and Historical.
   Consisting of "Physical Geology," which sets forth the Leading Principles of the Science; and "Historical Geology," which treats of the Mineral and Organic Conditions of the Earth at each successive epoch. By RALPH TATE, F.G.S. **4/6**

Electro-Metallurgy,
   Practically Treated. By ALEXANDER WATT. Ninth Edition, enlarged and revised, including the most Recent Processes . . . **3/6**

MECHANICAL ENGINEERING.

The Workman's Manual of Engineering Drawing.
By JOHN MAXTON, Instructor in Engineering Drawing, Royal Naval College, Greenwich. Seventh Edition. 300 Plates and Diagrams . **3/6**

Fuels: Solid, Liquid, and Gaseous.
Their Analysis and Valuation. For the Use of Chemists and Engineers. By H. J. PHILLIPS, F.C.S., formerly Analytical and Consulting Chemist to the Great Eastern Railway. Second Edition, Revised . . . **2/0**

Fuel, Its Combustion and Economy.
Consisting of an Abridgment of "A Treatise on the Combustion of Coal and the Prevention of Smoke." By C. W. WILLIAMS, A.I.C.E. With Extensive Additions by D. K. CLARK, M. Inst. C.E. Third Edition . **3/6**

The Boilermaker's Assistant
in Drawing, Templating, and Calculating Boiler Work, &c. By J. COURTNEY, Practical Boilermaker. Edited by D. K. CLARK, C.E. . **2/0**

The Boiler-Maker's Ready Reckoner,
with Examples of Practical Geometry and Templating for the Use of Platers, Smiths, and Riveters. By JOHN COURTNEY. Edited by D. K. CLARK, M.I.C.E. Second Edition, revised, with Additions . . **4/0**

** *The last two Works in One Volume, half-bound, entitled* "THE BOILER-MAKER'S READY-RECKONER AND ASSISTANT." By J. COURTNEY and D. K. CLARK. *Price* 7s.

Steam Boilers:
Their Construction and Management. By R. ARMSTRONG, C.E. Illustrated **1/6**

Steam and Machinery Management.
A Guide to the Arrangement and Economical Management of Machinery. By M. POWIS BALE, M. Inst. M.E. **2/6**

Steam and the Steam Engine,
Stationary and Portable. Being an Extension of the Treatise on the Steam Engine of Mr. J. SEWELL. By D. K. CLARK, C.E. Third Edition **3/6**

The Steam Engine,
A Treatise on the Mathematical Theory of, with Rules and Examples for Practical Men. By T. BAKER, C.E. **1/6**

The Steam Engine.
By Dr. LARDNER. Illustrated **1/6**

Locomotive Engines,
By G. D. DEMPSEY, C.E. With large Additions treating of the Modern Locomotive, by D. K. CLARK, M. Inst. C.E. **3/0**

Locomotive Engine-Driving.
A Practical Manual for Engineers in charge of Locomotive Engines. By MICHAEL REYNOLDS. Eighth Edition. 3s. 6d. limp; cloth boards **4/6**

Stationary Engine-Driving.
A Practical Manual for Engineers in charge of Stationary Engines. By MICHAEL REYNOLDS. Fourth Edition. 3s. 6d. limp; cloth boards . **4/6**

The Smithy and Forge.
Including the Farrier's Art and Coach Smithing. By W. J. E. CRANE. Second Edition, revised **2/6**

Modern Workshop Practice,
As applied to Marine, Land, and Locomotive Engines, Floating Docks, Dredging Machines, Bridges, Ship-building, &c. By J. G. WINTON. Fourth Edition, Illustrated **3/6**

Mechanical Engineering.
Comprising Metallurgy, Moulding, Casting, Forging, Tools, Workshop Machinery, Mechanical Manipulation, Manufacture of the Steam Engine, &c. By FRANCIS CAMPIN, C.E. Third Edition . . . **2/6**

Details of Machinery.
Comprising Instructions for the Execution of various Works in Iron in the Fitting-Shop, Foundry, and Boiler-Yard. By FRANCIS CAMPIN, C.E. **3/0**

Elementary Engineering:
 A Manual for Young Marine Engineers and Apprentices. In the Form of Questions and Answers on Metals, Alloys, Strength of Materials, &c. By J. S. BREWER. Second Edition 2/0

Power in Motion:
 Horse-power Motion, Toothed-Wheel Gearing, Long and Short Driving Bands, Angular Forces, &c. By JAMES ARMOUR, C.E. Third Edition 2/0

Iron and Heat,
 Exhibiting the Principles concerned in the Construction of Iron Beams, Pillars, and Girders. By J. ARMOUR, C.E. 2/6

Practical Mechanism,
 And Machine Tools. By T. BAKER, C.E. With Remarks on Tools and Machinery, by J. NASMYTH, C.E. 2/6

Mechanics:
 Being a concise Exposition of the General Principles of Mechanical Science, and their Applications. By CHARLES TOMLINSON, F.R.S. . . 1/6

Cranes (The Construction of),
 And other Machinery for Raising Heavy Bodies for the Erection of Buildings, &c. By JOSEPH GLYNN, F.R.S. 1/6

NAVIGATION, SHIPBUILDING, ETC.

The Sailor's Sea Book:
 A Rudimentary Treatise on Navigation. By JAMES GREENWOOD, B.A. With numerous Woodcuts and Coloured Plates. New and enlarged Edition. By W. H. ROSSER 2/6

Practical Navigation.
 Consisting of THE SAILOR'S SEA-BOOK, by JAMES GREENWOOD and W. H. ROSSER; together with Mathematical and Nautical Tables for the Working of the Problems, by HENRY LAW, C.E., and Prof. J. R. YOUNG . 7/0

Navigation and Nautical Astronomy,
 In Theory and Practice. By Prof. J. R. YOUNG. New Edition . 2/6

Mathematical Tables,
 For Trigonometrical, Astronomical, and Nautical Calculations; to which is prefixed a Treatise on Logarithms. By H. LAW, C.E. Together with a Series of Tables for Navigation and Nautical Astronomy. By Professor J. R. YOUNG. New Edition 4/0

Masting, Mast-Making, and Rigging of Ships.
 Also Tables of Spars, Rigging, Blocks; Chain, Wire, and Hemp Ropes, &c., relative to every class of vessels. By ROBERT KIPPING, N.A. . 2/0

Sails and Sail-Making.
 With Draughting, and the Centre of Effort of the Sails. By ROBERT KIPPING, N.A. 2/6

Marine Engines and Steam Vessels.
 By R. MURRAY, C.E. Eighth Edition, thoroughly Revised, with Additions by the Author and by GEORGE CARLISLE, C.E. . . . 4/6

Iron Ship-Building.
 With Practical Examples. By JOHN GRANTHAM. Fifth Edition . 4/0

Naval Architecture:
 An Exposition of Elementary Principles. By JAMES PEAKE . . 3/6

Ships for Ocean and River Service,
 Principles of the Construction of. By HAKON A. SOMMERFELDT . 1/6

Atlas of Engravings
 To Illustrate the above. Twelve large folding Plates. Royal 4to, cloth 7/6

The Forms of Ships and Boats.
 By W. BLAND. Seventh Edition, revised, with numerous Illustrations and Models 1/6

ARCHITECTURE AND THE BUILDING ARTS.

Constructional Iron and Steel Work,
as applied to Public, Private, and Domestic Buildings. By FRANCIS CAMPIN, C.E. 3/6

Building Estates:
A Treatise on the Development, Sale, Purchase, and Management of Building Land. By F. MAITLAND. Second Edition, revised . 2/0

The Science of Building:
An Elementary Treatise on the Principles of Construction. By E. WYNDHAM TARN, M.A. Lond. Third Edition, revised and enlarged . 3/6

The Art of Building:
General Principles of Construction, Strength, and Use of Materials, Working Drawings, Specifications, &c. By EDWARD DOBSON, M.R.I.B.A. . 2/0

A Book on Building,
Civil and Ecclesiastical. By Sir EDMUND BECKETT, Q.C. (Lord GRIMTHORPE). Second Edition 4/6

Dwelling-Houses (The Erection of),
Illustrated by a Perspective View, Plans, and Sections of a Pair of Villas, with Specification, Quantities, and Estimates. By S. H. BROOKS, Architect 2/6

Cottage Building.
By C. BRUCE ALLEN. Eleventh Edition, with Chapter on Economic Cottages for Allotments, by E. E. ALLEN, C.E. 2/0

Acoustics in Relation to Architecture and Building:
The Laws of Sound as applied to the Arrangement of Buildings. By Professor T. ROGER SMITH, F.R.I.B.A. New Edition, Revised . . 1/6

The Rudiments of Practical Bricklaying.
General Principles of Bricklaying; Arch Drawing, Cutting, and Setting; Pointing; Paving, Tiling, &c. By ADAM HAMMOND. With 68 Woodcuts 1/6

The Art of Practical Brick Cutting and Setting.
By ADAM HAMMOND. With 90 Engravings 1/6

Brickwork:
A Practical Treatise, embodying the General and Higher Principles of Bricklaying, Cutting and Setting; with the Application of Geometry to Roof Tiling, &c. By F. WALKER 1/6

Bricks and Tiles,
Rudimentary Treatise on the Manufacture of; containing an Outline of the Principles of Brickmaking. By E. DOBSON, M.R.I.B.A. Additions by C. TOMLINSON, F.R.S. Illustrated 3/0

The Practical Brick and Tile Book.
Comprising: BRICK AND TILE MAKING, by E. DOBSON, A.I.C.E.; Practical BRICKLAYING, by A. HAMMOND; BRICKWORK, by F. WALKER. 550 pp. with 270 Illustrations, strongly half-bound . . . 6/0

Carpentry and Joinery—
THE ELEMENTARY PRINCIPLES OF CARPENTRY. Chiefly composed from the Standard Work of THOMAS TREDGOLD, C.E. With Additions, and TREATISE ON JOINERY, by E. W. TARN, M.A. Fifth Edition, Revised . . 3/6

Carpentry and Joinery—Atlas
Of 35 Plates to accompany and illustrate the foregoing book. With Descriptive Letterpress. 4to. 6/0

WEALE'S SCIENTIFIC AND TECHNICAL SERIES.

A Practical Treatise on Handrailing;
Showing New and Simple Methods. By Geo. Collings. Second Edition. Revised, including a Treatise on Stairbuilding. With Plates . **2/6**

Circular Work in Carpentry and Joinery.
A Practical Treatise on Circular Work of Single and Double Curvature. By George Collings. Second Edition **2/6**

Roof Carpentry:
Practical Lessons in the Framing of Wood Roofs. For the Use of Working Carpenters. By Geo. Collings **2/0**

The Construction of Roofs of Wood and Iron;
Deduced chiefly from the Works of Robison, Tredgold, and Humber. By E. Wyndham Tarn, M.A., Architect. Second Edition, revised . **1/6**

The Joints Made and Used by Builders.
By Wyvill J. Christy, Architect. With 160 Woodcuts . . **3/0**

Shoring
And Its Application: A Handbook for the Use of Students. By George H. Blagrove. With 31 Illustrations **1/6**

The Timber Importer's, Timber Merchant's, and Builder's Standard Guide.
By R. E. Grandy **2/0**

Plumbing:
A Text-Book to the Practice of the Art or Craft of the Plumber. With Chapters upon House Drainage and Ventilation. By Wm. Paton Buchan. Sixth Edition, revised and enlarged, with 380 Illustrations . . **3/6**

Ventilation:
A Text Book to the Practice of the Art of Ventilating Buildings. By W. P. Buchan, R.P., Author of "Plumbing," &c. With 170 Illustrations **3/6**

The Practical Plasterer:
A Compendium of Plain and Ornamental Plaster Work. By W. Kemp **2/0**

House Painting, Graining, Marbling, & Sign Writing.
With a Course of Elementary Drawing, and a Collection of Useful Receipts. By Ellis A. Davidson. Sixth Edition. Coloured Plates . . **5/0**
*** *The above, in cloth boards, strongly bound, 6s.*

A Grammar of Colouring,
Applied to Decorative Painting and the Arts. By George Field. New Edition, enlarged, by Ellis A. Davidson. With Coloured Plates . **3/0**

Elementary Decoration
As applied to Dwelling Houses, &c. By James W. Facey. Illustrated **2/0**

Practical House Decoration.
A Guide to the Art of Ornamental Painting, the Arrangement of Colours in Apartments, and the Principles of Decorative Design. By James W. Facey. **2/6**
*** *The last two Works in One handsome Vol., half-bound, entitled "*House Decoration, Elementary and Practical,*" price 5s.*

Warming and Ventilation
Of Domestic and Public Buildings, Mines, Lighthouses, Ships, &c. By Charles Tomlinson, F.R.S. **3/0**

Portland Cement for Users.
By Henry Faija, A.M. Inst. C.E. Third Edition, Corrected . **2/0**

Limes, Cements, Mortars, Concretes, Mastics, Plastering, &c.
By G. R. Burnell, C.E. Thirteenth Edition **1/6**

Masonry and Stone-Cutting.
The Principles of Masonic Projection and their application to Construction. By EDWARD DOBSON, M.R.I.B.A. 2/6

Arches, Piers, Buttresses, &c.:
Experimental Essays on the Principles of Construction. By W. BLAND. 1/6

Quantities and Measurements,
In Bricklayers', Masons', Plasterers', Plumbers', Painters', Paperhangers', Gilders', Smiths', Carpenters' and Joiners' Work. By A. C. BEATON 1/6

The Complete Measurer:
Setting forth the Measurement of Boards, Glass, Timber and Stone. By R. HORTON. Fifth Edition 4/0

** *The above, strongly bound in leather, price 5s.*

Light:
An Introduction to the Science of Optics. Designed for the Use of Students of Architecture, Engineering, and other Applied Sciences. By E. WYNDHAM TARN, M.A., Author of "The Science of Building," &c. . 1/6

Hints to Young Architects.
By GEORGE WIGHTWICK, Architect. Fifth Edition, revised and enlarged by G. HUSKISSON GUILLAUME, Architect 3/6

Architecture—Orders:
The Orders and their Æsthetic Principles. By W. H. LEEDS. Illustrated. 1/6

Architecture—Styles:
The History and Description of the Styles of Architecture of Various Countries, from the Earliest to the Present Period. By T. TALBOT BURY, F.R.I.B.A. Illustrated 2/0

** ORDERS AND STYLES OF ARCHITECTURE, *in One Vol., 3s. 6d.*

Architecture—Design:
The Principles of Design in Architecture, as deducible from Nature and exemplified in the Works of the Greek and Gothic Architects. By EDW. LACY GARBETT, Architect. Illustrated 2/6

** *The three preceding Works in One handsome Vol., half bound, entitled* "MODERN ARCHITECTURE," *price 6s.*

Perspective for Beginners.
Adapted to Young Students and Amateurs in Architecture, Painting, &c. By GEORGE PYNE 2/0

Architectural Modelling in Paper.
By T. A. RICHARDSON. With Illustrations, engraved by O. JEWITT 1/6

Glass Staining, and the Art of Painting on Glass.
From the German of Dr. GESSERT and EMANUEL OTTO FROMBERG. With an Appendix on THE ART OF ENAMELLING 2/6

Vitruvius—The Architecture of.
In Ten Books. Translated from the Latin by JOSEPH GWILT, F.S.A., F.R.A.S. With 23 Plates 5/0

N.B.—This is the only Edition of VITRUVIUS *procurable at a moderate price*

Grecian Architecture,
An Inquiry into the Principles of Beauty in. With an Historical View of the Rise and Progress of the Art in Greece. By the EARL OF ABERDEEN 1/0

** *The two preceding Works in One handsome Vol., half bound, entitled* "ANCIENT ARCHITECTURE," *price 6s.*

INDUSTRIAL AND USEFUL ARTS.

Cements, Pastes, Glues, and Gums.
A Practical Guide to the Manufacture and Application of the various Agglutinants required for Workshop, Laboratory, or Office Use. With upwards of 900 Recipes and Formulæ. By H. C. STANDAGE . . 2/0

Clocks and Watches, and Bells,
A Rudimentary Treatise on. By Sir EDMUND BECKETT, Q.C. (Lord GRIMTHORPE). Seventh Edition 4/6

The Goldsmith's Handbook.
Containing full Instructions in the Art of Alloying, Melting, Reducing, Colouring, Collecting and Refining, Recovery of Waste, Solders, Enamels, &c., &c. By GEORGE E. GEE. Third Edition, enlarged . . . 3/0

The Silversmith's Handbook,
On the same plan as the GOLDSMITH'S HANDBOOK. By GEORGE E. GEE. Second Edition, Revised 3/0
*** *The last two Works, in One handsome Vol., half-bound, 7s.*

The Hall-Marking of Jewellery.
Comprising an account of all the different Assay Towns of the United Kingdom; with the Stamps and Laws relating to the Standards and Hall-Marks at the various Assay Offices. By GEORGE E. GEE . . 3/0

Practical Organ Building.
By W. E. DICKSON, M.A. Second Edition, Revised, with Additions 2/6

Coach-Building:
A Practical Treatise. By JAMES W. BURGESS. With 57 Illustrations 2/6

The Brass Founder's Manual:
Instructions for Modelling, Pattern Making, Moulding, Turning, &c. By W. GRAHAM 2/0

The Sheet-Metal Worker's Guide.
A Practical Handbook for Tinsmiths, Coppersmiths, Zincworkers, &c., with 46 Diagrams. By W. J. E. CRANE. Second Edition, revised . 1/6

Sewing Machinery:
Its Construction, History, &c. With full Technical Directions for Adjusting, &c. By J. W. URQUHART, C.E. 2/0

Gas Fitting:
A Practical Handbook. By JOHN BLACK. Second Edition, Enlarged. With 130 Illustrations 2/6

Construction of Door Locks.
From the Papers of A. C. HOBBS. Edited by CHARLES TOMLINSON, F.R.S. With a Note upon IRON SAFES by ROBERT MALLET. Illustrated . 2/6

The Model Locomotive Engineer, Fireman, and Engine-Boy.
Comprising an Historical Notice of the Pioneer Locomotive Engines and their Inventors. By MICHAEL REYNOLDS. Second Edition. With numerous Illustrations, and Portrait of George Stephenson . . 3/6

The Art of Letter Painting made Easy.
By J. G. BADENOCH. With 12 full-page Engravings of Examples . 1/6

The Art of Boot and Shoemaking.
Including Measurement, Last-fitting, Cutting-out, Closing and Making. By JOHN BEDFORD LENO. With numerous Illustrations. Third Edition 2/0

Mechanical Dentistry:
A Practical Treatise on the Construction of the Various Kinds of Artificial Dentures. By CHARLES HUNTER. Third Edition, revised . . 3/0

Wood Engraving:
A Practical and Easy Introduction to the Art. By W. N. BROWN . 1/6

Laundry Management.
A Handbook for Use in Private and Public Laundries. Including Accounts of Modern Machinery and Appliances. By the EDITOR of "The Laundry Journal." With numerous Illustrations. Second Edition . . 2/0

AGRICULTURE, GARDENING, ETC.

Draining and Embanking:
A Practical Treatise. By Prof. JOHN SCOTT. With 68 Illustrations **1/6**

Irrigation and Water Supply:
A Practical Treatise on Water Meadows, Sewage Irrigation, Warping, &c.; on the Construction of Wells, Ponds, Reservoirs, &c. By Prof. JOHN SCOTT. With 34 Illustrations **1/6**

Farm Roads, Fences, and Gates:
A Practical Treatise on the Roads, Tramways, and Waterways of the Farm; the Principles of Enclosures; and the different kinds of Fences, Gates, and Stiles. By Prof. JOHN SCOTT. With 75 Illustrations . **1/6**

Farm Buildings:
A Practical Treatise on the Buildings necessary for various kinds of Farms, their Arrangement and Construction, with Plans and Estimates. By Prof. JOHN SCOTT. With 105 Illustrations **2/0**

Barn Implements and Machines:
Treating of the Application of Power and Machines used in the Threshing-barn, Stockyard, Dairy, &c. By Prof. J. SCOTT. With 123 Illustrations. **2/0**

Field Implements and Machines:
With Principles and Details of Construction and Points of Excellence, their Management, &c. By Prof. JOHN SCOTT. With 138 Illustrations **2/0**

Agricultural Surveying:
A Treatise on Land Surveying, Levelling, and Setting-out; with Directions for Valuing Estates. By Prof. J. SCOTT. With 62 Illustrations . **1/6**

Farm Engineering.
By Professor JOHN SCOTT. Comprising the above Seven Volumes in One, 1,150 pages, and over 600 Illustrations. Half-bound . . . **12/0**

Outlines of Farm Management.
Treating of the General Work of the Farm; Stock; Contract Work; Labour, &c. By R. SCOTT BURN **2/6**

Outlines of Landed Estates Management.
Treating of the Varieties of Lands, Methods of Farming, Setting-out of Farms, Roads, Fences, Gates, Drainage, &c. By R. SCOTT BURN . **2/6**
*** *The above Two Vols. in One, handsomely half-bound, price* **6s.**

Soils, Manures, and Crops.
(Vol. I. OUTLINES OF MODERN FARMING.) By R. SCOTT BURN . **2/0**

Farming and Farming Economy.
(Vol. II. OUTLINES OF MODERN FARMING.) By R. SCOTT BURN **3/0**

Stock: Cattle, Sheep, and Horses.
(Vol. III. OUTLINES OF MODERN FARMING.) By R. SCOTT BURN **2/6**

Dairy, Pigs, and Poultry.
(Vol. IV. OUTLINES OF MODERN FARMING.) By R. SCOTT BURN **2/0**

Utilization of Sewage, Irrigation, and Reclamation of Waste Land.
(Vol. V. OUTLINES OF MODERN FARMING.) By R. SCOTT BURN . **2/6**

Outlines of Modern Farming.
By R. SCOTT BURN. Consisting of the above Five Volumes in One, 1,250 pp., profusely Illustrated, half-bound **12/0**

Book-keeping for Farmers and Estate Owners.
A Practical Treatise, presenting, in Three Plans, a System adapted for all classes of Farms. By J. M. WOODMAN. Third Edition, revised . **2/6**

Ready Reckoner for the Admeasurement of Land.
By A. ARMAN. Third Edition, revised and extended by C. NORRIS **2/0**

Miller's, Corn Merchant's, and Farmer's Ready Reckoner.
Second Edition, revised, with a Price List of Modern Flour Mill Machinery, by W. S. HUTTON, C.E. **2/0**

The Hay and Straw Measurer.
New Tables for the Use of Auctioneers, Valuers, Farmers, Hay and Straw Dealers, &c. By JOHN STEELE. **2/0**

Meat Production.
A Manual for Producers, Distributors, and Consumers of Butchers' Meat. By JOHN EWART **2/6**

Sheep:
The History, Structure, Economy, and Diseases of. By W. C. SPOONER, M.R.V.S. Fifth Edition, with fine Engravings. **3/6**

Market and Kitchen Gardening.
By C. W. SHAW, late Editor of "Gardening Illustrated". . . **3/0**

Kitchen Gardening Made Easy.
Showing the best means of Cultivating every known Vegetable and Herb, &c., with directions for management all the year round. By GEORGE M. F. GLENNY. Illustrated **1/6**

Cottage Gardening:
Or Flowers, Fruits, and Vegetables for Small Gardens. By E. HOBDAY.
1/6

Garden Receipts.
Edited by CHARLES W. QUIN **1/6**

Fruit Trees,
The Scientific and Profitable Culture of. From the French of M. DU BREUIL. Fourth Edition, carefuly Revised by GEORGE GLENNY. With 187 Woodcuts **3/6**

The Tree Planter and Plant Propagator:
With numerous Illustrations of Grafting, Layering, Budding, Implements, Houses, Pits, &c. By SAMUEL WOOD **2/0**

The Tree Pruner:
A Practical Manual on the Pruning of Fruit Trees, Shrubs, Climbers, and Flowering Plants. With numerous Illustrations. By SAMUEL WOOD **1/6**

*** *The above Two Vols. in One, handsomely half-bound, price* 3*s.* 6*d.*

The Art of Grafting and Budding.
By CHARLES BALTET. With Illustrations **2/6**

MATHEMATICS, ARITHMETIC, ETC.

Descriptive Geometry,
An Elementary Treatise on; with a Theory of Shadows and of Perspective, extracted from the French of G. MONGE. To which is added a Description of the Principles and Practice of Isometrical Projection. By J. F. HEATHER, M.A. With 14 Plates **2/0**

Practical Plane Geometry:
Giving the Simplest Modes of Constructing Figures contained in one Plane and Geometrical Construction of the Ground. By J. F. HEATHER, M.A. With 215 Woodcuts **2/0**

Analytical Geometry and Conic Sections,
A Rudimentary Treatise on. By JAMES HANN. A New Edition, re-written and enlarged by Professor J. R. YOUNG **2/0**

Euclid (The Elements of).
With many Additional Propositions and Explanatory Notes; to which is prefixed an Introductory Essay on Logic. By HENRY LAW, C.E. . **2/6**

*** *Sold also separately, viz:—*
   Euclid. The First Three Books. By HENRY LAW, C.E. . . **1/6**
   Euclid. Books 4, 5, 6, 11, 12. By HENRY LAW, C.E. . . **1/6**

Plane Trigonometry,
The Elements of. By JAMES HANN **1/6**

Spherical Trigonometry,
The Elements of. By JAMES HANN. Revised by CHARLES H. DOWLING, C.E. **1/0**
*** *Or with "The Elements of Plane Trigonometry," in One Volume, 2s. 6d.*

Differential Calculus,
Elements of the. By W. S. B. WOOLHOUSE, F.R.A.S., &c. . **1/6**

Integral Calculus.
By HOMERSHAM COX, B.A. **1/0**

Algebra,
The Elements of. By JAMES HADDON, M.A. With Appendix, containing Miscellaneous Investigations, and a collection of Problems . **2/0**

A Key and Companion to the Above.
An extensive repository of Solved Examples and Problems in Algebra. By J. R. YOUNG. **1/6**

Commercial Book-keeping.
With Commercial Phrases and Forms in English, French, Italian, and German. By JAMES HADDON, M.A. **1/6**

Arithmetic,
A Rudimentary Treatise on. With full Explanations of its Theoretical Principles, and numerous Examples for Practice. For the Use of School and for Self-Instruction. By J. R. YOUNG, late Professor of Mathematics in Belfast College. Eleventh Edition **1/6**

A Key to the Above.
By J. R. YOUNG **1/6**

Equational Arithmetic,
Applied to Questions of Interest, Annuities, Life Assurance, and General Commerce; with various Tables by which all Calculations may be greatly facilitated. By W. HIPSLEY **2/0**

Arithmetic,
Rudimentary, for the Use of Schools and Self-Instruction. By JAMES HADDON, M.A. Revised by ABRAHAM ARMAN . . . **1/6**

A Key to the Above.
A. ARMAN **1/6**

Mathematical Instruments:
Their Construction, Adjustment, Testing, and Use concisely explained. By J. F. HEATHER, M.A., of the Royal Military Academy, Woolwich. Fourteenth Edition, Revised, with Additions, by A. T. WALMISLEY, M.I.C.E. Original Edition, in 1 vol., Illustrated . . . **2/0**

*** *In ordering the above, be careful to say "Original Edition," or give the number in the Series (32), to distinguish it from the Enlarged Edition in 3 vols. (Nos. 168-9-70).*

Drawing and Measuring Instruments.
Including—I. Instruments employed in Geometrical and Mechanical Drawing, and in the Construction, Copying, and Measurement of Maps and Plans. II. Instruments used for the purposes of Accurate Measurement, and for Arithmetical Computations. By J. F. HEATHER, M.A. . **1/6**

Optical Instruments.
Including (more especially) Telescopes, Microscopes, and Apparatus for producing copies of Maps and Plans by Photography. By J. F. HEATHER, M.A. Illustrated **1/6**

Surveying and Astronomical Instruments.
Including—I. Instruments used for Determining the Geometrical Features of a portion of Ground. II. Instruments employed in Astronomical Observations. By J. F. HEATHER, M.A. Illustrated . . **1/6**

*** *The above three volumes form an enlargement of the Author's original work, "Mathematical Instruments:" price 2s. (See No. 32 in the Series.)*

Mathematical Instruments:
Their Construction, Adjustment, Testing and Use. Comprising Drawing, Measuring, Optical, Surveying, and Astronomical Instruments. By J. F. HEATHER, M.A. Enlarged Edition, for the most part entirely re-written. The Three Parts as above, in One thick Volume . . . **4/6**

The Slide Rule, and How to Use It.
Containing full, easy, and simple Instructions to perform all Business Calculations with unexampled rapidity and accuracy. By CHARLES HOARE, C.E. With a Slide Rule, in tuck of cover. Fifth Edition . . **2/6**

Logarithms.
With Mathematical Tables for Trigonometrical, Astronomical, and Nautical Calculations. By HENRY LAW, C.E. Revised Edition. (Forming part of the above work.) **3/0**

Compound Interest and Annuities (Theory of).
With Tables of Logarithms for the more Difficult Computations of Interest, Discount, Annuities, &c., in all their Applications and Uses for Mercantile and State Purposes. By FEDOR THOMAN, Paris. Fourth Edition . **4/0**

Mathematical Tables,
For Trigonometrical, Astronomical, and Nautical Calculations; to which is prefixed a Treatise on Logarithms. By H. LAW, C.E. Together with a Series of Tables for Navigation and Nautical Astronomy. By Professor J. R. YOUNG. New Edition **4/0**

Mathematics,
As applied to the Constructive Arts. By FRANCIS CAMPIN, C.E., &c. Second Edition **3/0**

Astronomy.
By the late Rev. ROBERT MAIN, F.R.S. Third Edition, revised and corrected to the Present Time. By W. T. LYNN, F.R.A.S. . . . **2/0**

Statics and Dynamics,
The Principles and Practice of. Embracing also a clear development of Hydrostatics, Hydrodynamics, and Central Forces. By T. BAKER, C.E. Fourth Edition **1/6**

BOOKS OF REFERENCE AND MISCELLANEOUS VOLUMES.

A Dictionary of Painters, and Handbook for Picture Amateurs.
Being a Guide for Visitors to Public and Private Picture Galleries, and for Art-Students, including Glossary of Terms, Sketch of Principal Schools of Painting, &c. By PHILIPPE DARYL, B.A. **2/6**

Painting Popularly Explained.
By T. J. GULLICK, Painter, and JOHN TIMBS, F.S.A. Including Fresco, Oil, Mosaic, Water Colour, Water-Glass, Tempera, Encaustic, Miniature, Painting on Ivory, Vellum, Pottery, Enamel, Glass, &c. Fifth Edition **5/0**

A Dictionary of Terms used in Architecture, Building, Engineering, Mining, Metallurgy, Archæology, the Fine Arts, &c.
By JOHN WEALE. Sixth Edition. Edited by ROBT. HUNT, F.R.S. Numerous Illustrations **5/0**

Music:
A Rudimentary and Practical Treatise. With numerous Examples. By CHARLES CHILD SPENCER **2/6**

Pianoforte,
The Art of Playing the. With numerous Exercises and Lessons. By CHARLES CHILD SPENCER **1/6**

The House Manager.
Being a Guide to Housekeeping, Practical Cookery, Pickling and Preserving, Household Work, Dairy Management, Cellarage of Wines, Home-brewing and Wine-making, Stable Economy, Gardening Operations, &c. By AN OLD HOUSEKEEPER **3/6**

Manual of Domestic Medicine.
By R. GOODING M.D. Intended as a Family Guide in all cases of Accident and Emergency Third Edition, carefully revised . . **2/0**

Management of Health.
A Manual of Home and Personal Hygiene. By Rev. JAMES BAIRD **1/0**

Natural Philosophy,
For the Use of Beginners. By CHARLES TOMLINSON, F.R.S. . . **1/6**

The Electric Telegraph,
Its History and Progress. With Descriptions of some of the Apparatus. By R. SABINE, C.E., F.S.A., &c. **3/0**

Handbook of Field Fortification.
By Major W. W. KNOLLYS, F.R.G.S. With 163 Woodcuts . . **3/0**

Logic,
Pure and Applied. By S. H. EMMENS. Third Edition . . . **1/6**

Locke on the Human Understanding,
Selections from. With Notes by S. H. EMMENS . . . **1/6**

The Compendious Calculator
(*Intuitive Calculations*). Or Easy and Concise Methods of Performing the various Arithmetical Operations required in Commercial and Business Transactions; together with Useful Tables, &c. By DANIEL O'GORMAN. Twenty-seventh Edition, carefully revised by C. NORRIS . . . **2/6**

Measures, Weights, and Moneys of all Nations.
With an Analysis of the Christian, Hebrew, and Mahometan Calendars. By W. S. B. WOOLHOUSE, F.R.A.S., F.S.S. Seventh Edition . **2/6**

Grammar of the English Tongue,
Spoken and Written. With an Introduction to the Study of Comparative Philology. By HYDE CLARKE, D.C.L. Fifth Edition. . . . **1/6**

Dictionary of the English Language,
As Spoken and Written. Containing above 100,000 Words. By HYDE CLARKE, D.C.L. **3/6**
  Complete with the GRAMMAR . . . **5/6**

Composition and Punctuation,
Familiarly Explained for those who have neglected the Study of Grammar. By JUSTIN BRENAN. 18th Edition **1/6**

French Grammar.
With Complete and Concise Rules on the Genders of French Nouns. By G. L. STRAUSS, Ph.D. **1/6**

French-English Dictionary.
Comprising a large number of New Terms used in Engineering, Mining, &c. By ALFRED ELWES **1/6**

English-French Dictionary.
By ALFRED ELWES **2/0**

French Dictionary.
The two Parts, as above, complete in One Vol. **3/0**
  ⁎⁎ *Or with the* GRAMMAR, **4/6**.

French and English Phrase Book.
Containing Introductory Lessons, with Translations, Vocabularies of Words, Collection of Phrases, and Easy Familiar Dialogues **1/6**

German Grammar.
Adapted for English Students, from Heyse's Theoretical and Practical Grammar, by Dr. G. L. STRAUSS **1/6**

German Triglot Dictionary.
By N. E. S. A. HAMILTON. Part I. German-French-English. Part II. English-German-French. Part III. French-German-English . . **3/0**

German Triglot Dictionary
(As above). Together with German Grammar in One Volume . . **5/0**

Italian Grammar
Arranged in Twenty Lessons, with Exercises. By ALFRED ELWES. **1/6**

Italian Triglot Dictionary,
Wherein the Genders of all the Italian and French Nouns are carefully noted down. By ALFRED ELWES. Vol. 1. Italian-English-French . **2/6**

Italian Triglot Dictionary.
By ALFRED ELWES. Vol. 2. English-French-Italian . . **2/6**

Italian Triglot Dictionary.
By ALFRED ELWES. Vol. 3. French-Italian-English . . **2/6**

Italian Triglot Dictionary
(As above). In One Vol. **7/6**

Spanish Grammar.
In a Simple and Practical Form. With Exercises. By ALFRED ELWES **1/6**

Spanish-English and English-Spanish Dictionary.
Including a large number of Technical Terms used in Mining, Engineering, &c., with the proper Accents and the Gender of every Noun. By ALFRED ELWES **4/0**
  ⁎⁎ *Or with the* GRAMMAR, **6/0**.

Portuguese Grammar,
In a Simple and Practical Form. With Exercises. By ALFRED ELWES **1/6**

Portuguese-English and English-Portuguese Dictionary.
Including a large number of Technical Terms used in Mining, Engineering, &c., with the proper Accents and the Gender of every Noun. By ALFRED ELWES. Third Edition, revised **5/0**

*** Or with the GRAMMAR, 7/0.*

Animal Physics,
Handbook of. By DIONYSIUS LARDNER, D.C.L. With 520 Illustrations. In One Vol. (732 pages), cloth boards **7/6**

*** Sold also in Two Parts, as follows:—*

ANIMAL PHYSICS. By Dr. LARDNER. Part I., Chapters I.—VII. **4/0**

ANIMAL PHYSICS. By Dr. LARDNER. Part II., Chapters VIII.—XVIII. **3/0**

September, 1896.

A CATALOGUE OF BOOKS
INCLUDING NEW AND STANDARD WORKS IN
ENGINEERING: CIVIL, MECHANICAL, AND MARINE;
ELECTRICITY AND ELECTRICAL ENGINEERING;
MINING, METALLURGY; ARCHITECTURE,
BUILDING, INDUSTRIAL AND DECORATIVE ARTS;
SCIENCE, TRADE AND MANUFACTURES;
AGRICULTURE, FARMING, GARDENING;
AUCTIONEERING, VALUING AND ESTATE AGENCY;
LAW AND MISCELLANEOUS.
PUBLISHED BY
CROSBY LOCKWOOD & SON.

MECHANICAL ENGINEERING, etc.

D. K. Clark's Pocket-Book for Mechanical Engineers.
THE MECHANICAL ENGINEER'S POCKET-BOOK OF TABLES, FORMULÆ, RULES AND DATA. A Handy Book of Reference for Daily Use in Engineering Practice. By D. KINNEAR CLARK, M.Inst.C.E., Author of "Railway Machinery," "Tramways," &c. Third Edition, Revised. Small 8vo, 700 pages, 6s. bound in flexible leather cover, rounded corners.

SUMMARY OF CONTENTS.
MATHEMATICAL TABLES.—MEASUREMENT OF SURFACES AND SOLIDS.—ENGLISH WEIGHTS AND MEASURES.—FRENCH METRIC WEIGHTS AND MEASURES.—FOREIGN WEIGHTS AND MEASURES.—MONEYS.—SPECIFIC GRAVITY, WEIGHT AND VOLUME.—MANUFACTURED METALS.—STEEL PIPES.—BOLTS AND NUTS.—SUNDRY ARTICLES IN WROUGHT AND CAST IRON, COPPER, BRASS, LEAD, TIN, ZINC.—STRENGTH OF MATERIALS.—STRENGTH OF TIMBER.—STRENGTH OF CAST IRON.—STRENGTH OF WROUGHT IRON.—STRENGTH OF STEEL.—TENSILE STRENGTH OF COPPER, LEAD, ETC.—RESISTANCE OF STONES AND OTHER BUILDING MATERIALS.—RIVETED JOINTS IN BOILER PLATES.—BOILER SHELLS.—WIRE ROPES AND HEMP ROPES.—CHAINS AND CHAIN CABLES.—FRAMING.—HARDNESS OF METALS, ALLOYS AND STONES.—LABOUR OF ANIMALS.—MECHANICAL PRINCIPLES.—GRAVITY AND FALL OF BODIES.—ACCELERATING AND RETARDING FORCES.—MILL GEARING, SHAFTING, ETC.—TRANSMISSION OF MOTIVE POWER.—HEAT.—COMBUSTION: FUELS.—WARMING, VENTILATION, COOKING STOVES.—STEAM.—STEAM ENGINES AND BOILERS.—RAILWAYS.—TRAMWAYS.—STEAM SHIPS.—PUMPING STEAM ENGINES AND PUMPS.—COAL GAS, GAS ENGINES, ETC.—AIR IN MOTION.—COMPRESSED AIR.—HOT AIR ENGINES.—WATER POWER.—SPEED OF CUTTING TOOLS.—COLOURS.—ELECTRICAL ENGINEERING.

*** OPINIONS OF THE PRESS.

"Mr Clark manifests what is an innate perception of what is likely to be useful in a pocket-book, and he is really unrivalled in the art of condensation. Very frequently we find the information on a given subject is supplied by giving a summary description of an experiment, and a statement of the results obtained. There is a very excellent steam table, occupying five and-a-half pages; and there are rules given for several calculations, which rules cannot be found in other pocket-books, as, for example, that on page 497, for getting at the quantity of water in the shape of priming in any known weight of steam. It is very difficult to hit upon any mechanical engineering subject concerning which this work supplies no information, and the excellent index at the end adds to its utility. In one word, it is an exceedingly handy and efficient tool, possessed of which the engineer will be saved many a wearisome calculation, or yet more wearisome hunt through various text-books and treatises. and, as such, we can heartily recommend it to our readers, who must not run away with the idea that Mr. Clark's Pocket-book is only Molesworth in another form. On the contrary, each contains what is not to be found in the other; and Mr. Clark takes more room and deals at more length with many subjects than Molesworth possibly could."—*The Engineer.*

"It would be found difficult to compress more matter within a similar compass, or produce a book of 650 pages which should be more compact or convenient for pocket reference. . . . Will be appreciated by mechanical engineers of all classes."—*Practical Engineer.*

"Just the kind of work that practical men require to have near to them."—*English Mechanic.*

MR. HUTTON'S PRACTICAL HANDBOOKS.

Handbook for Works' Managers.

THE WORKS' MANAGER'S HANDBOOK OF MODERN RULES, TABLES, AND DATA. For Engineers, Millwrights, and Boiler Makers; Tool Makers, Machinists, and Metal Workers; Iron and Brass Founders, &c. By W. S. HUTTON, Civil and Mechanical Engineer, Author of "The Practical Engineer's Handbook." Fifth Edition, carefully Revised, with Additions. In One handsome Volume, medium 8vo, price 15s. strongly bound. [*Just published.*

☞ *The Author having compiled Rules and Data for his own use in a great variety of modern engineering work, and having found his notes extremely useful, decided to publish them—revised to date—believing that a practical work, suited to the* DAILY REQUIREMENTS OF MODERN ENGINEERS, *would be favourably received.*

In the Fourth Edition the First Section has been re-written and improved by the addition of numerous Illustrations and new matter relating to STEAM ENGINES *and* GAS ENGINES. *The Second Section has been enlarged and Illustrated, and throughout the book a great number of emendations and alterations have been made, with the object of rendering the book more generally useful.*

*** OPINIONS OF THE PRESS.

"The author treats every subject from the point of view of one who has collected workshop notes for application in workshop practice, rather than from the theoretical or literary aspect. The volume contains a great deal of that kind of information which is gained only by practical experience, and is seldom written in books."—*Engineer.*

"The volume is an exceedingly useful one, brimful with engineers' notes, memoranda, and rules, and well worthy of being on every mechanical engineer's bookshelf."—*Mechanical World.*

"The information is precisely that likely to be required in practice. . . . The work forms a desirable addition to the library not only of the works' manager, but of anyone connected with general engineering."—*Mining Journal.*

"A formidable mass of facts and figures, readily accessible through an elaborate index . . . Such a volume will be found absolutely necessary as a book of reference in all sorts of 'works' connected with the metal trades."—*Ryland's Iron Trades Circular.*

"Brimful of useful information, stated in a concise form, Mr. Hutton's books have met a pressing want among engineers. The book must prove extremely useful to every practical man possessing a copy."—*Practical Engineer.*

New Manual for Practical Engineers.

THE PRACTICAL ENGINEER'S HAND-BOOK. Comprising a Treatise on Modern Engines and Boilers: Marine, Locomotive and Stationary. And containing a large collection of Rules and Practical Data relating to recent Practice in Designing and Constructing all kinds of Engines, Boilers, and other Engineering work. The whole constituting a comprehensive Key to the Board of Trade and other Examinations for Certificates of Competency in Modern Mechanical Engineering. By WALTER S. HUTTON, Civil and Mechanical Engineer, Author of "The Works' Manager's Handbook for Engineers," &c. With upwards of 370 Illustrations. Fifth Edition, Revised, with Additions. Medium 8vo, nearly 500 pp., price 18s. Strongly bound. [*Just published.*

☞ *This work is designed as a companion to the Author's* "WORKS' MANAGER'S HAND-BOOK." *It possesses many new and original features, and contains, like its predecessor, a quantity of matter not originally intended for publication, but collected by the author for his own use in the construction of a great variety of* MODERN ENGINEERING WORK.

The information is given in a condensed and concise form, and is illustrated by upwards of 370 Woodcuts; and comprises a quantity of tabulated matter of great value to all engaged in designing, constructing, or estimating for ENGINES, BOILERS, *and* OTHER ENGINEERING WORK.

*** OPINIONS OF THE PRESS.

"We have kept it at hand for several weeks, referring to it as occasion arose, and we have not on a single occasion consulted its pages without finding the information of which we were in quest."—*Athenæum.*

"A thoroughly good practical handbook, which no engineer can go through without learning something that will be of service to him."—*Marine Engineer.*

"An excellent book of reference for engineers, and a valuable text-book for students of engineering."—*Scotsman.*

"This valuable manual embodies the results and experience of the leading authorities on mechanical engineering."—*Building News.*

"The author has collected together a surprising quantity of rules and practical data, and has shown much judgment in the selections he has made. . . . There is no doubt that this book is one of the most useful of its kind published, and will be a very popular compendium."—*Engineer.*

"A mass of information, set down in simple language, and in such a form that it can be easily referred to at any time. The matter is uniformly good and well chosen and is greatly elucidated by the illustrations. The book will find its way on to most engineers' shelves, where it will rank as one of the most useful books of reference."—*Practical Engineer.*

"Full of useful information and should be found on the office shelf of all practical engineers."—*English Mechanic.*

MECHANICAL ENGINEERING, etc. 3

MR. HUTTON'S PRACTICAL HANDBOOKS—continued.

Practical Treatise on Modern Steam-Boilers.
STEAM-BOILER CONSTRUCTION. A Practical Handbook for Engineers, Boiler-Makers, and Steam Users. Containing a large Collection of Rules and Data relating to Recent Practice in the Design, Construction, and Working of all Kinds of Stationary, Locomotive, and Marine Steam-Boilers. By WALTER S. HUTTON, Civil and Mechanical Engineer, Author of "The Works' Manager's Handbook," "The Practical Engineer's Handbook," &c. With upwards of 300 Illustrations. Second Edition. Medium 8vo, 18s. cloth.

☞ *This work is issued in continuation of the Series of Handbooks written by the Author, viz:—"THE WORKS' MANAGER'S HANDBOOK" and "THE PRACTICAL ENGINEER'S HANDBOOK," which are so highly appreciated by Engineers for the practical nature of their information; and is consequently written in the same style as those works.*

The Author believes that the concentration, in a convenient form for easy reference, of such a large amount of thoroughly practical information on Steam-Boilers, will be of considerable service to those for whom it is intended, and he trusts the book may be deemed worthy of as favourable a reception as has been accorded to its predecessors.

*** OPINIONS OF THE PRESS.

"Every detail, both in boiler design and management, is clearly laid before the reader. The volume shows that boiler construction has been reduced to the condition of one of the most exact sciences; and such a book is of the utmost value to the *fin de siècle* Engineer and Works Manager."—*Marine Engineer.*

"There has long been room for a modern handbook on steam boilers; there is not that room now, because Mr. Hutton has filled it. It is a thoroughly practical book for those who are occupied in the construction, design, selection, or use of boilers."—*Engineer.*

"The book is of so important and comprehensive a character that it must find its way into the libraries of everyone interested in boiler using or boiler manufacture if they wish to be thoroughly informed. We strongly recommend the book for the intrinsic value of its contents."—*Machinery Market.*

"The value of this book can hardly be over-estimated The author's rules, formulæ &c., are all very fresh, and it is impossible to turn to the work and not find what you want. No practical engineer should be without it."—*Colliery Guardian.*

Hutton's "Modernised Templeton."
THE PRACTICAL MECHANICS' WORKSHOP COMPANION. Comprising a great variety of the most useful Rules and Formulæ in Mechanical Science, with numerous Tables of Practical Data and Calculated Results for Facilitating Mechanical Operations. By WILLIAM TEMPLETON, Author of "The Engineer's Practical Assistant," &c. &c. Seventeenth Edition, Revised, Modernised, and considerably Enlarged by WALTER S. HUTTON, C.E., Author of "The Works' Manager's Handbook," "The Practical Engineer's Handbook," &c. Fcap. 8vo, nearly 500 pp., with 8 Plates and upwards of 250 Illustrative Diagrams, 6s., strongly bound for workshop or pocket wear and tear.

*** OPINIONS OF THE PRESS.

"In its modernised form Hutton's 'Templeton' should have a wide sale, for it contains much valuable information which the mechanic will often find of use, and not a few tables and notes which he might look for in vain in other works. This modernised edition will be appreciated by all who have learned to value the original editions of 'Templeton.'"—*English Mechanic.*

"It has met with great success in the engineering workshop, as we can testify; and there are a great many men who, in a great measure, owe their rise in life to this little book."—*Building News.*

"This familiar text-book—well known to all mechanics and engineers—is of essential service to the every-day requirements of engineers, millwrights, and the various trades connected with engineering and building. The new modernised edition is worth its weight in gold."—*Building News.* (Second Notice.)

"This well-known and largely-used book contains information, brought up to date, of the sort so useful to the foreman and draughtsman. So much fresh information has been introduced as to constitute it practically a new book. It will be largely used in the office and workshop."—*Mechanical World.*

"The publishers wisely entrusted the task of revision of this popular, valuable, and useful book to Mr. Hutton, than whom a more competent man they could not have found."—*Iron.*

Templeton's Engineer's and Machinist's Assistant.
THE ENGINEER'S, MILLWRIGHT'S, and MACHINIST'S PRACTICAL ASSISTANT. A collection of Useful Tables, Rules and Data. By WILLIAM TEMPLETON. 7th Edition, with Additions. 18mo, 2s. 6d. cloth.

*** OPINIONS OF THE PRESS.

"Occupies a foremost place among books of this kind. A more suitable present to an apprentice to any of the mechanical trades could not possibly be made."—*Building News.*

"A deservedly popular work. It should be in the 'drawer' of every mechanic."—*English Mechanic.*

Foley's Office Reference Book for Mechanical Engineers.

THE MECHANICAL ENGINEER'S REFERENCE BOOK, for Machine and Boiler Construction. In Two Parts. Part I. GENERAL ENGINEERING DATA. Part II. BOILER CONSTRUCTION. With 51 Plates and numerous Illustrations. By NELSON FOLEY, M.I.N.A. Second Edition, Revised throughout and much Enlarged. Folio, £3 3s. net half-bound.

[Just published.

SUMMARY OF CONTENTS.

PART I.

MEASURES.—CIRCUMFERENCES AND AREAS, &c., SQUARES, CUBES, FOURTH POWERS.—SQUARE AND CUBE ROOTS.—SURFACE OF TUBES—RECIPROCALS.—LOGARITHMS.—MENSURATION.—SPECIFIC GRAVITIES AND WEIGHTS.—WORK AND POWER.—HEAT.—COMBUSTION.—EXPANSION AND CONTRACTION.—EXPANSION OF GASES.—STEAM.—STATIC FORCES.—GRAVITATION AND ATTRACTION.—MOTION AND COMPUTATION OF RESULTING FORCES.—ACCUMULATED WORK.—CENTRE AND RADIUS OF GYRATION.—MOMENT OF INERTIA.—CENTRE OF OSCILLATION.—ELECTRICITY.—STRENGTH OF MATERIALS.—ELASTICITY.—TEST SHEETS OF METALS.—FRICTION.—TRANSMISSION OF POWER.—FLOW OF LIQUIDS.—FLOW OF GASES.—AIR PUMPS, SURFACE CONDENSERS, &c.—SPEED OF STEAMSHIPS.—PROPELLERS.—CUTTING TOOLS.—FLANGES.—COPPER SHEETS AND TUBES.—SCREWS, NUTS, BOLT HEADS, &c.—VARIOUS RECIPES AND MISCELLANEOUS MATTER.

WITH DIAGRAMS FOR VALVE-GEAR, BELTING AND ROPES, DISCHARGE AND SUCTION PIPES, SCREW PROPELLERS, AND COPPER PIPES.

PART II.

TREATING OF, POWER OF BOILERS.—USEFUL RATIOS.—NOTES ON CONSTRUCTION.—CYLINDRICAL BOILER SHELLS.—CIRCULAR FURNACES.—FLAT PLATES.—STAYS.—GIRDERS.—SCREWS.—HYDRAULIC TESTS.—RIVETING.—BOILER SETTING, CHIMNEYS, AND MOUNTINGS.—FUELS, &c.—EXAMPLES OF BOILERS AND SPEEDS OF STEAMSHIPS.—NOMINAL AND NORMAL HORSE POWER.

WITH DIAGRAMS FOR ALL BOILER CALCULATIONS AND DRAWINGS OF MANY VARIETIES OF BOILERS.

*** OPINIONS OF THE PRESS.

"The book is one which every mechanical engineer may, with advantage to himself add to his library."—*Industries.*

"Mr. Foley is well fitted to compile such a work. . . . The diagrams are a great feature of the work. . . . Regarding the whole work, it may be very fairly stated that Mr. Foley has produced a volume which will undoubtedly fulfil the desire of the author and become indispensable to all mechanical engineers."—*Marine Engineer.*

"We have carefully examined this work, and pronounce it a most excellent reference book for the use of marine engineers."—*Journal of American Society of Naval Engineers.*

"A veritable monument of industry on the part of Mr. Foley, who has succeeded in producing what is simply invaluable to the engineering profession."—*Steamship.*

Coal and Speed Tables.

A POCKET BOOK OF COAL AND SPEED TABLES, for Engineers and Steam-users. By NELSON FOLEY, Author of "The Mechanical Engineer's Reference Book." Pocket-size, 3s. 6d. cloth.

"These tables are designed to meet the requirements of every-day use; they are of sufficient scope for most practical purposes, and may be commended to engineers and users of steam."—*Iron.*

"This pocket-book well merits the attention of the practical engineer. Mr. Foley has compiled a very useful set of tables, the information contained in which is frequently required by engineers, coal consumers and users of steam."—*Iron and Coal Trades Review.*

Steam Engine.

TEXT-BOOK ON THE STEAM ENGINE. With a Supplement on Gas Engines, and PART II. ON HEAT ENGINES. By T. M. GOODEVE, M.A., Barrister-at-Law, Professor of Mechanics at the Royal College of Science, London; Author of "The Principles of Mechanics," "The Elements of Mechanism," &c. Twelfth Edition, Enlarged. With numerous Illustrations. Crown 8vo, 6s. cloth.

"Professor Goodeve has given us a treatise on the steam engine which will bear comparison with anything written by Huxley or Maxwell, and we can award it no higher praise."—*Engineer.*

"Mr. Goodeve's text-book is a work of which every young engineer should possess himself."—*Mining Journal.*

MECHANICAL ENGINEERING.

Gas Engines.

ON GAS-ENGINES. With Appendix describing a Recent Engine with Tube Igniter. By T. M. GOODEVE, M.A. Crown 8vo, 2s. 6d. cloth. [*Just published.*

"Like all Mr. Goodeve's writings, the present is no exception in point of general excellence It is a valuable little volume."—*Mechanical World.*

Steam Engine Design.

A HANDBOOK ON THE STEAM ENGINE, with especial Reference to Small and Medium-sized Engines. For the Use of Engine-Makers, Mechanical Draughtsmen, Engineering Students and Users of Steam Power. By HERMAN HAEDER, C.E. English Edition, Re-edited by the Author from the Second German Edition, and Translated, with considerable Additions and Alterations, by H. H. P. POWLES, A.M.I.C.E., M.I.M.E. With nearly 1,100 Illustrations. Crown 8vo, 9s. cloth.

"A perfect encyclopædia of the steam engine and its details, and one which must take a permanent place in English drawing-offices and workshops."—*A Foreman Pattern-maker.*

"This is an excellent book, and should be in the hands of all who are interested in the construction and design of medium-sized stationary engines. . . . A careful study of its contents and the arrangement of the sections leads to the conclusion that there is probably no other book like it in this country. The volume aims at showing the results of practical experience, and it certainly may claim a complete achievement of this idea."—*Nature.*

"There can be no question as to its value. We cordially commend it to all concerned in the design and construction of the steam engine."—*Mechanical World.*

Steam Boilers.

A TREATISE ON STEAM BOILERS: Their Strength, Construction, and Economical Working. By ROBERT WILSON, C.E. Fifth Edition. 12mo, 6s. cloth.

"The best treatise that has ever been published on steam boilers."—*Engineer.*

"The author shows himself perfect master of his subject, and we heartily recommend all employing steam power to possess themselves of the work."—*Ryland's Iron Trade Circular.*

Boiler Chimneys.

BOILER AND FACTORY CHIMNEYS; Their Draught-Power and Stability. With a Chapter on Lightning Conductors. By ROBERT WILSON, A.I.C.E., Author of "A Treatise on Steam Boilers," &c. Second Edition. Crown 8vo, 3s. 6d. cloth.

"A valuable contribution to the literature of scientific building."—*The Builder.*

Boiler Making.

THE BOILER-MAKER'S READY RECKONER & ASSISTANT. With Examples of Practical Geometry and Templating, for the Use of Platers, Smiths and Riveters. By JOHN COURTNEY, Edited by D. K. CLARK, M.I.C.E. Third Edition, 480 pp., with 140 Illusts. Fcap. 8vo, 7s. half-bound.

"No workman or apprentice should be without this book."—*Iron Trade Circular.*

Locomotive Engine Development.

THE LOCOMOTIVE ENGINE AND ITS DEVELOPMENT. A Popular Treatise on the Gradual Improvements made in Railway Engines between 1803 and 1896. By CLEMENT E. STRETTON, C.E., Author of "Safe Railway Working," &c. Fifth Edition, Revised and Enlarged. With 120 Illustrations. Crown 8vo, 3s. 6d. cloth gilt. [*Just published.*

"Students of railway history and all who are interested in the evolution of the modern locomotive will find much to attract and entertain in this volume."—*The Times.*

"The author of this work is well known to the railway world, and no one probably has a better knowledge of the history and development of the locomotive. The volume before us should be of value to all connected with the railway system of this country."—*Nature.*

Estimating for Engineering Work, &c.

ENGINEERING ESTIMATES, COSTS AND ACCOUNTS: A Guide to Commercial Engineering. With numerous Examples of Estimates and Costs of Millwright Work, Miscellaneous Productions, Steam Engines and Steam Boilers; and a Section on the Preparation of Costs Accounts. By A GENERAL MANAGER. Demy 8vo, 12s. cloth.

"This is an excellent and very useful book, covering subject-matter in constant requisition to every factory and workshop. . . . The book is invaluable, not only to the young engineer, but also to the estimate department of every works."—*Builder.*

"We accord the work unqualified praise. The information is given in a plain, straightforward manner, and bears throughout evidence of the intimate practical acquaintance of the author with every phase of commercial engineering."—*Mechanical World.*

Fire Engineering.

FIRES, FIRE-ENGINES, AND FIRE-BRIGADES. With a History of Fire-Engines, their Construction, Use, and Management; Remarks on Fire-Proof Buildings, and the Preservation of Life from Fire; Statistics of the Fire Appliances in English Towns; Foreign Fire Systems; Hints on Fire-Brigades, &c. &c. By CHARLES F. T. YOUNG, C.E. With numerous Illustrations. 544 pp., demy 8vo, £1 4s. cloth.

"To those interested in the subject of fires and fire apparatus, we most heartily commend this book. It is the only English work we now have upon the subject."—*Engineering*.

Boilermaking.

PLATING AND BOILERMAKING: A Practical Handbook for Workshop Operations. By JOSEPH G. HORNER, A.M.I.M.E. (Foreman Pattern-Maker), Author of "Pattern Making," &c. 380 pages, with 338 Illustrations. Crown 8vo, 7s. 6d. cloth. [*Just published*.

"The latest production from the pen of this writer is characterised by that evidence of close acquaintance with workshop methods which will render the book exceedingly acceptable to the practical hand. We have no hesitation in commending the work as a serviceable and practical handbook on a subject which has not hitherto received much attention from those qualified to deal with it in a satisfactory manner."—*Mechanical World*.

Engineering Construction.

PATTERN-MAKING: *A Practical Treatise*, embracing the Main Types of Engineering Construction, and including Gearing, both Hand and Machine made, Engine Work, Sheaves and Pulleys, Pipes and Columns, Screws, Machine Parts, Pumps and Cocks, the Moulding of Patterns in Loam and Greensand, &c., together with the methods of Estimating the weight of Castings; to which is added an Appendix of Tables for Workshop Reference. By JOSEPH G. HORNER, A.M.I.M.E. (Foreman Pattern-Maker). Second Edition, thoroughly Revised and much Enlarged. With upwards of 450 Illustrations. Crown 8vo, 7s. 6d. cloth. [*Just published*.

"A well-written technical guide, evidently written by a man who understands and has practised what he has written about. . . . We cordially recommend it to engineering students, young journeymen, and others desirous of being initiated into the mysteries of pattern-making."—*Builder*.

"More than 450 illustrations help to explain the text, which is, however, always clear and explicit, thus rendering the work an excellent *vade mecum* for the apprentice who desires to become master of his trade."—*English Mechanic*.

Dictionary of Mechanical Engineering Terms.

LOCKWOOD'S DICTIONARY OF TERMS USED IN THE PRACTICE OF MECHANICAL ENGINEERING, embracing those current in the Drawing Office, Pattern Shop, Foundry, Fitting, Turning, Smith's and Boiler Shops, &c. &c. Comprising upwards of 6,000 Definitions. Edited by JOSEPH G. HORNER, A.M.I.M.E. (Foreman Pattern-Maker), Author of "Pattern Making." Second Edition, Revised. Crown 8vo, 7s. 6d. cloth.

"Just the sort of handy dictionary required by the various trades engaged in mechanical engineering. The practical engineering pupil will find the book of great value in his studies, and every foreman engineer and mechanic should have a copy."—*Building News*.

"Not merely a dictionary, but, to a certain extent, also a most valuable guide. It strikes us as a happy idea to combine with a definition of the phrase useful information on the subject of which t treats."—*Machinery Market*.

Mill Gearing.

TOOTHED GEARING: A Practical Handbook for Offices and Workshops. By JOSEPH G. HORNER, A.M.I.M.E. (Foreman Pattern-Maker), Author of "Pattern Making," &c. With 184 Illustrations. Crown 8vo, 6s. cloth. [*Just published*.

SUMMARY OF CONTENTS.

CHAP. I. PRINCIPLES.—II. FORMATION OF TOOTH PROFILES.—III. PROPORTIONS OF TEETH.—IV. METHODS OF MAKING TOOTH FORMS.—V. INVOLUTE TEETH.—VI. SOME SPECIAL TOOTH FORMS.—VII. BEVEL WHEELS.—VIII. SCREW GEARS.—IX. WORM GEARS.—X. HELICAL WHEELS.—XI. SKEW BEVELS.—XII. VARIABLE AND OTHER GEARS.—XIII. DIAMETRICAL PITCH.—XIV. THE ODONTOGRAPH.—XV. PATTERN GEARS.—XVI. MACHINE MOULDING GEARS.—XVII. MACHINE CUT GEARS.—XVIII. PROPORTION OF WHEELS.

"We must give the book our unqualified praise for its thoroughness of treatment, and we can heartily recommend it to all interested as the most practical book on the subject yet written."—*Mechanical World*.

Stone-working Machinery.

STONE-WORKING MACHINERY, and the Rapid and Economical Conversion of Stone. With Hints on the Arrangement and Management of Stone Works. By M. POWIS BALE, M.I.M.E. With Illusts. Crown 8vo, 9s.

"The book should be in the hands of every mason or student of stone-work."—*Colliery Guardian.*

"A capital handbook for all who manipulate stone for building or ornamental purposes."—*Machinery Market.*

Pump Construction and Management.

PUMPS AND PUMPING : A Handbook for Pump Users. Being Notes on Selection, Construction and Management. By M. POWIS BALE, M.I.M.E., Author of " Woodworking Machinery," " Saw Mills," &c. Second Edition, Revised. Crown 8vo, 2s. 6d. cloth.

"The matter is set forth as concisely as possible. In fact, condensation rather than diffuseness has been the author's aim throughout; yet he does not seem to have omitted anything likely to be of use."—*Journal of Gas Lighting.*

"Thoroughly practical and simply and clearly written."—*Glasgow Herald.*

Milling Machinery, etc.

MILLING MACHINES AND PROCESSES : A Practical Treatise on Shaping Metals by Rotary Cutters, including Information on Making and Grinding the Cutters. By PAUL N. HASLUCK, Author of " Lathe-work," " Handybooks for Handicrafts," &c. With upwards of 300 Engravings, including numerous Drawings by the Author. Large crown 8vo, 352 pages, 12s. 6d. cloth.

"A new departure in engineering literature. . . . We can recommend this work to all interested nmilling machines ; it is what it professes to be—a practical treatise."—*Engineer.*

"A capital and reliable book, which will no doubt be of considerable service, both to those who are already acquainted with the process as well as to those who contemplate its adoption."—*Industries.*

Turning.

LATHE-WORK : A Practical Treatise on the Tools, Appliances, and Processes employed in the Art of Turning. By PAUL N. HASLUCK. Fifth Edition, Revised and Enlarged Cr. 8vo, 5s. cloth.

"Written by a man who knows, not only how work ought to be done, but who also knows how to do it, and how to convey his knowledge to others. To all turners this book would be valuable."—*Engineering.*

"We can safely recommend the work to young engineers. To the amateur it will simply be invaluable. To the student it will convey a great deal of useful information."—*Engineer.*

Screw-Cutting.

SCREW THREADS : And Methods of Producing Them. With Numerous Tables, and complete directions for using Screw-Cutting Lathes. By PAUL N. HASLUCK, Author of " Lathe-Work," &c. With Seventy-four Illustrations. Third Edition, Revised and Enlarged. Waistcoat-pocket size, 1s. 6d. cloth.

"Full of useful information, hints and practical criticism. Taps, dies and screwing-tools generally are illustrated and their action described."—*Mechanical World.*

"It is a complete compendium of all the details of the screw cutting lathe ; in fact a *multum in parvo* on all the subjects it treats upon."—*Carpenter and Builder.*

Smith's Tables for Mechanics, etc.

TABLES, MEMORANDA, AND CALCULATED RESULTS, FOR MECHANICS, ENGINEERS, ARCHITECTS, BUILDERS, etc. Selected and Arranged by FRANCIS SMITH. Sixth Edition, Revised, including ELECTRICAL TABLES, FORMULÆ, and MEMORANDA. Waistcoat-pocket size, 1s. 6d. limp leather. [*Just published.*

"It would, perhaps, be as difficult to make a small pocket-book selection of notes and formulæ to suit ALL engineers as it would be to make a universal medicine ; but Mr. Smith's waistcoat-pocket collection may be looked upon as a successful attempt."—*Engineer.*

"The best example we have ever seen of 270 pages of useful matter packed into the dimensions of a card-case."—*Building News.* "A veritable pocket treasury of knowledge."—*Iron.*

French-English Glossary for Engineers, etc.

A POCKET GLOSSARY of TECHNICAL TERMS : ENGLISH-FRENCH, FRENCH-ENGLISH ; with Tables suitable for the Architectural, Engineering, Manufacturing and Nautical Professions. By JOHN JAMES FLETCHER, Engineer and Surveyor. Second Edition, Revised and Enlarged, 200 pp. Waistcoat-pocket size, 1s. 6d. limp leather.

"It is a very great advantage for readers and correspondents in France and England to have so large a number of the words relating to engineering and manufacturers collected in a liliputian volume. The little book will be useful both to students and travellers."—*Architect.*

"The glossary of terms is very complete, and many of the tables are new and well arranged. We cordially commend the book."—*Mechanical World.*

Year-Book of Engineering Formulæ, &c.

THE ENGINEER'S YEAR-BOOK FOR 1896. Comprising Formulæ, Rules, Tables, Data and Memoranda in Civil, Mechanical, Electrical, Marine and Mine Engineering. By H. R. KEMPE, A.M. Inst.C.E., M.I.E.E., Technical Officer of the Engineer-in-Chief's Office, General Post Office, London, Author of "A Handbook of Electrical Testing," "The Electrical Engineer's Pocket-Book," &c. With 800 Illustrations, specially Engraved for the work. Crown 8vo, 670 pages, 8s. leather. [*Just published.*

"Represents an enormous quantity of work and forms a desirable book of reference."—*The Engineer.*

"The book is distinctly in advance of most similar publications in this country."—*Engineering.*

"This valuable and well-designed book of reference meets the demands of all descriptions of engineers."—*Saturday Review.*

"Teems with up-to-date information in every branch of engineering and construction."—*Building News.*

"The needs of the engineering profession could hardly be supplied in a more admirable, complete and convenient form. To say that it more than sustains all comparisons is praise of the highest sort, and that may justly be said of it."—*Mining Journal.*

"There is certainly room for the new comer, which supplies explanations and directions, as well as formulæ and tables. It deserves to become one of the most successful of the technical annuals."—*Architect.*

"Brings together with great skill all the technical information which an engineer has to use day by day. It is in every way admirably equipped, and is sure to prove successful."—*Scotsman.*

"The up-to-dateness of Mr. Kempe's compilation is a quality that will not be lost on the busy people for whom the work is intended."—*Glasgow Herald.*

Portable Engines.

THE PORTABLE ENGINE; ITS CONSTRUCTION AND MANAGEMENT. A Practical Manual for Owners and Users of Steam Engines generally. By WILLIAM DYSON WANSBROUGH. With 90 Illustrations. Crown 8vo, 3s. 6d. cloth.

"This is a work of value to those who use steam machinery. . . . Should be read by everyone who has a steam engine, on a farm or elsewhere."—*Mark Lane Express.*

"We cordially commend this work to buyers and owners of steam engines, and to those who have to do with their construction or use."—*Timber Trades Journal.*

"Such a general knowledge of the steam engine as Mr. Wansbrough furnishes to the reader should be acquired by all intelligent owners and others who use the steam engine."—*Building News.*

"An excellent text-book of this useful form of engine. 'The Hints to Purchasers' contain a good deal of commonsense and practical wisdom."—*English Mechanic.*

Iron and Steel.

"IRON AND STEEL": *A Work for the Forge, Foundry, Factory, and Office.* Containing ready, useful, and trustworthy Information for Iron-masters and their Stock-takers; Managers of Bar, Rail, Plate, and Sheet Rolling Mills; Iron and Metal Founders; Iron Ship and Bridge Builders; Mechanical, Mining, and Consulting Engineers; Architects, Contractors, Builders, and Professional Draughtsmen. By CHARLES HOARE, Author of "The Slide Rule," &c. Ninth Edition. 32mo, 6s. leather.

"For comprehensiveness the book has not its equal."—*Iron.*

"One of the best of the pocket books."—*English Mechanic.*

"We cordially recommend this book to those engaged in considering the details of all kinds of iron and steel works."—*Naval Science.*

Elementary Mechanics.

CONDENSED MECHANICS. A Selection of Formulæ, Rules, Tables, and Data for the Use of Engineering Students, Science Classes, &c. In Accordance with the Requirements of the Science and Art Department. By W. G. CRAWFORD HUGHES, A.M.I.C.E. Crown 8vo, 2s 6d. cloth.

"The book is well fitted for those who are either confronted with practical problems in their work, or are preparing for examination and wish to refresh their knowledge by going through their formulæ again."—*Marine Engineer.*

"It is well arranged, and meets the wants of those for whom it is intended."—*Railway News.*

Steam.

THE SAFE USE OF STEAM. Containing Rules for Unprofessional Steam-users. By an ENGINEER. Seventh Edition. Sewed, 6d.

"If steam-users would but learn this little book by heart, boiler explosions would become sensations by their rarity."—*English Mechanic.*

Warming.

HEATING BY HOT WATER; with Information and Suggestions on the best Methods of Heating Public, Private and Horticultural Buildings. By WALTER JONES. Second Edition. With 96 Illustrations. Crown 8vo, 2s. 6d. net.

"We confidently recommend all interested in heating by hot water to secure a copy of this valuable little treatise."—*The Plumber and Decorator.*

CIVIL ENGINEERING, SURVEYING, etc.

Light Railways.

LIGHT RAILWAYS FOR THE UNITED KINGDOM, INDIA, AND THE COLONIES: A Practical Handbook setting forth the Principles on which Light Railways should be Constructed, Worked and Financed; and detailing the cost of Construction, Equipment, Revenue, and Working Expences of Local Railways already established in the above-mentioned Countries, and in Belgium, France, Switzerland, &c. By JOHN CHARLES MACKAY, F.G.S., A.M.Inst.C.E. Illustrated with 40 Photographic Plates and other Diagrams. Medium 8vo, 15s. cloth. [*Just published.*
"Exactly what has been long wanted, and sure to have a wide sale."—*Railway News.*

Water Supply and Water-Works.

THE WATER SUPPLY OF TOWNS AND THE CONSTRUCTION OF WATER-WORKS: A Practical Treatise for the Use of Engineers and Students of Engineering. By W. K. BURTON, A.M.Inst.C E., Professor of Sanitary Engineering in the Imperial University, Tokyo, Japan, and Consulting Engineer to the Tokyo Water-works. With an Appendix on The Effects of Earthquakes on Waterworks, by JOHN MILNE, F.R.S., Professor of Mining in the Imperial University of Japan. With numerous Plates and Illustrations. Super-royal 8vo, 25s. buckram. [*Just published.*
" The whole art of waterworks construction is dealt with in a clear and comprehensive fashion in this handsome volume. . . . Mr. Burton's practical treatise shows in all its sections the fruit of independent study and individual experience. It is largely based upon his own practice in the branch of engineering of which it treats, and with such a basis a treatise can scarcely fail to be suggestive and useful."—*Saturday Review.*
'Professor Burton's book is sure of a warm welcome among engineers. It is written in clear and vigorous language and forms an exhaustive treatise on a branch of engineering the claims of which it would be difficult to over-estimate."—*Scotsman.*
" The subjects seem to us to be ably discussed, with a practical aim to meet the requirements of all its probable readers. The volume is well got up, and the illustrations are excellent."
The Lancet

Water Supply of Cities and Towns.

A COMPREHENSIVE TREATISE on the WATER-SUPPLY OF CITIES AND TOWNS. By WILLIAM HUMBER, A-M.Inst.C.E., and M. Inst. M.E., Author of "Cast and Wrought Iron Bridge Construction," &c. &c. Illustrated with 50 Double Plates, 1 Single Plate, Coloured Frontispiece, and upwards of 250 Woodcuts, and containing 400 pages of Text. Imp. 4to. £6 6s. elegantly and substantially half-bound in morocco.
" The most systematic and valuable work upon water supply hitherto produced in English or in any other language. . . . Mr. Humber's work is characterised almost throughout by an exhaustiveness much more distinctive of French and German than of English technical treatises. —*Engineer.*
"We can congratulate Mr. Humber on having been able to give so large an amount of information on a subject so important as the water supply of cities and towns. The plates, fifty in number, are mostly drawings of executed works, and alone would have commanded the attention of every engineer whose practice may lie in this branch of the profession."—*Builder.*

Water Supply.

RURAL WATER SUPPLY: A Practical Handbook on the Supply of Water and Construction of Waterworks for small Country Districts. By ALLAN GREENWELL, A.M.I.C.E., and W. T. CURRY, A.M.I.C.E., F.G.S. With Illustrations. Crown 8vo, 5s. cloth. [*Just published.*
"We conscientiously recommend it as a very useful book for those concerned in obtaining water for small districts, giving a great deal of practical information in a small compass."—*Builder.*
" The volume contains valuable information upon all matters connected with water supply. . . It is full of details on points which are continually before waterworks' engineers."—*Nature.*

Hydraulic Tables.

HYDRAULIC TABLES, CO-EFFICIENTS, and FORMULÆ for finding the Discharge of Water from Orifices, Notches, Weirs, Pipes, and Rivers. By JOHN NEVILLE, Civil Engineer, M.R.I.A. Third Ed., carefully Revised, with considerable Additions. Numerous Illusts. Cr. 8vo, 14s. cloth.
"Alike valuable to students and engineers in practice; its study will prevent the annoyance of avoidable failures, and assist them to select the readiest means of successfully carrying out any given work connected with hydraulic engineering."—*Mining Journal.*
" It is, of all English books on the subject, the one nearest to completeness,".—*Architect.*

Hydraulics.

HYDRAULIC MANUAL. Consisting of Working Tables and Explanatory Text. Intended as a Guide in Hydraulic Calculations and Field Operations. By LOWIS D'A. JACKSON, Author of "Aid to Survey Practice," "Modern Metrology," &c. Fourth Edition, Enlarged. Large cr. 8vo, 16s. cl.

"The author has had a wide experience in hydraulic engineering and has been a careful observer of the facts which have come under his notice, and from the great mass of material at his command he has constructed a manual which may be accepted as a trustworthy guide to this branch of the engineer's profession. We can heartily recommend this volume to all who desire to be acquainted with the latest development of this important subject."—*Engineering*.

"The standard-work in this department of mechanics."—*Scotsman.*

"The most useful feature of this work is its freedom from what is superannuated, and its thorough adoption of recent experiments; the text is, in fact, in great part a short account of the great modern experiments."—*Nature.*

Water Storage, Conveyance, and Utilisation.

WATER ENGINEERING: A Practical Treatise on the Measurement, Storage, Conveyance, and Utilisation of Water for the Supply of Towns, for Mill Power, and for other Purposes. By CHARLES SLAGG, A.M.Inst.C.E., Author of "Sanitary Work in the Smaller Towns, and in Villages," &c. Second Edition. With numerous Illustrations. Crown 8vo, 7s. 6d. cloth.

"As a small practical treatise on the water supply of towns, and on some applications of water-power, the work is in many respects excellent."—*Engineering.*

"The author has collated the results deduced from the experiments of the most eminent authorities, and has presented them in a compact and practical form, accompanied by very clear and detailed explanations. . . . The application of water as a motive power is treated very carefully and exhaustively."—*Builder.*

"For anyone who desires to begin the study of hydraulics with a consideration of the practical applications of the science there is no better guide."—*Architect.*

Drainage.

ON THE DRAINAGE OF LANDS, TOWNS, AND BUILDINGS. By G. D. DEMPSEY, C.E., Author of "The Practical Railway Engineer," &c. Revised, with large Additions on RECENT PRACTICE IN DRAINAGE ENGINEERING, by D. KINNEAR CLARK, M.Inst.C.E. Author of "Tramways: Their Construction and Working," "A Manual of Rules, Tables, and Data for Mechanical Engineers," &c. Third Edition. Small crown 8vo, 4s. 6d. cloth. [*Just published.*

"The new matter added to Mr. Dempsey's excellent work is characterised by the comprehensive grasp and accuracy of detail for which the name of Mr. D. K. Clark is a sufficient voucher."—*Athenæum.*

"As a work on recent practice in drainage engineering, the book is to be commended to all who are making that branch of engineering science their special study."—*Iron.*

"A comprehensive manual on drainage engineering, and a useful introduction to the student."—*Building News.*

River Engineering.

RIVER BARS: The Causes of their Formation, and their Treatment by "Induced Tidal Scour;" with a Description of the Successful Reduction by this Method of the Bar at Dublin. By I. J. MANN, Assist. Eng. to the Dublin Port and Docks Board. Royal 8vo, 7s. 6d. cloth.

"We recommend all interested in harbour works—and, indeed, those concerned in the improvements of rivers generally—to read Mr. Mann's interesting work on the treatment of river bars."—*Engineer.*

Tramways and their Working.

TRAMWAYS: THEIR CONSTRUCTION AND WORKING. Embracing a Comprehensive History of the System; with an exhaustive Analysis of the various Modes of Traction, including Horse-Power, Steam, Cable Traction, Electric Traction, &c.; a Description of the Varieties of Rolling Stock; and ample Details of Cost and Working Expenses. New Edition, Thoroughly Revised, and Including the Progress recently made in Tramway Construction, &c. &c. By D. KINNEAR CLARK. M.Inst.C.E. With numerous Illustrations and Folding Plates. In One Volume, 8vo, 780 pages, price 28s., bound in buckram. [*Just published.*

"All interested in tramways must refer to it, as all railway engineers have turned to the author's work 'Railway Machinery.'"—*Engineer.*

"An exhaustive and practical work on tramways, in which the history of this kind of locomotion, and a description and cost of the various modes of laying tramways, are to be found. —*Building News.*

"The best form of rails, the best mode of construction, and the best mechanical appliance are so fairly indicated in the work under review, that any engineer about to construct a tramway will be enabled at once to obtain the practical information which will be of most service to him. —*Athenæum.*

CIVIL ENGINEERING, SURVEYING, etc.

Student's Text-Book on Surveying.
PRACTICAL SURVEYING: A Text-Book for Students preparing for Examination or for Survey-work in the Colonies. By GEORGE W. USILL, A.M.I.C.E., Author of "The Statistics of the Water Supply of Great Britain." With Four Lithographic Plates and upwards of 330 Illustrations. Fourth Edition, Revised, including Tables of Natural Sines, Tangents, Secants, &c. Crown 8vo, 7s. 6d. cloth; or, on THIN PAPER, bound in limp leather, gilt edges, rounded corners, for pocket use, 12s. 6d.

"The best forms of instruments are described as to their construction, uses and modes of employment, and there are innumerable hints on work and equipment such; as the author, in his experience as surveyor, draughtsman, and teacher, has found necessary, and which the student in his inexperience will find most serviceable."—*Engineer.*

"The latest treatise in the English language on surveying, and we have no hesitation in saying that the student will find it a better guide than any of its predecessors . . . Deserves to be recognised as the first book which should be put in the hands of a pupil of Civil Engineering, and every gentleman of education who sets out for the Colonies would find it well to have a copy."—*Architect.*

Survey Practice.
AID TO SURVEY PRACTICE, for Reference in Surveying, Levelling, and Setting-out; and in Route Surveys of Travellers by Land and Sea. With Tables, Illustrations, and Records. By LOWIS D'A. JACKSON, A.M.I.C.E., Author of "Hydraulic Manual," "Modern Metrology," &c. Second Edition, Enlarged. Large crown 8vo, 12s. 6d. cloth.

"A valuable *vade-mecum* for the surveyor. We can recommend this book as containing an admirable supplement to the teaching of the accomplished surveyor."—*Athenæum.*

"As a text-book we should advise all surveyors to place it in their libraries, and study well the matured instructions afforded in its pages."—*Colliery Guardian.*

"The author brings to his work a fortunate union of theory and practical experience which, aided by a clear and lucid style of writing, renders the book a very useful one."—*Builder.*

Surveying, Land and Marine.
LAND AND MARINE SURVEYING, in Reference to the Preparation of Plans for Roads and Railways; Canals, Rivers, Towns' Water Supplies; Docks and Harbours. With Description and Use of Surveying Instruments. By W. D. HASKOLL, C.E., Author of "Bridge and Viaduct Construction," &c. Second Edition, Revised, with Additions. Large cr. 8vo, 9s. cl.

"This book must prove of great value to the student. We have no hesitation in recommending it, feeling assured that it will more than repay a careful study."—*Mechanical World.*

"A most useful and well arranged book. We can strongly recommend it as a carefully-written and valuable text-book. It enjoys a well-deserved repute among surveyors."—*Builder.*

"This volume cannot fail to prove of the utmost practical utility. It may be safely recommended to all students who aspire to become clean and expert surveyors."—*Mining Journal.*

Field-Book for Engineers.
THE ENGINEER'S, MINING SURVEYOR'S, AND CONTRACTOR'S FIELD-BOOK. Consisting of a Series of Tables, with Rules, Explanations of Systems, and use of Theodolite for Traverse Surveying and Plotting the Work with minute accuracy by means of Straight Edge and Set Square only; Levelling with the Theodolite, Casting-out and Reducing Levels to Datum, and Plotting Sections in the ordinary manner; setting-out Curves with the Theodolite by Tangential Angles and Multiples, with Right and Left-hand Readings of the Instrument: Setting-out Curves without Theodolite, on the System of Tangential Angles by sets of Tangents and Offsets; and Earthwork Tables to 80 feet deep, calculated for every 6 inches in depth. By W. D. HASKOLL, C.E. Fourth Edition. Crown 8vo, 12s. cloth.

"The book is very handy; the separate tables of sines and tangents to every minute will make useful for many other purposes, the genuine traverse tables existing all the same."—*Athenæum.*

"Every person engaged in engineering field operations will estimate the importance of such a work and the amount of valuable time which will be saved by reference to a set of reliable tables prepared with the accuracy and fulness of those given in this volume."—*Railway News.*

Levelling.
A TREATISE ON THE PRINCIPLES AND PRACTICE OF LEVELLING. Showing its Application to purposes of Railway and Civil Engineering, in the Construction of Roads; with Mr. TELFORD's Rules for the same. By FREDERICK W. SIMMS, F.G.S., M.Inst.C.E. Seventh Edition, with the addition of LAW's Practical Examples for Setting-out Railway Curves, and TRAUTWINE's Field Practice of Laying-out Circular Curves. With 7 Plates and numerous Woodcuts. 8vo, 8s. 6d. cloth. *** TRAUTWINE on Curves may be had separate, 5s.

"The text-book on levelling in most of our engineering schools and colleges. . . . The publishers have rendered a substantial service to the profession, especially to the younger members, by bringing out the present edition of Mr. Simms's useful book."—*Engineer.*

Trigonometrical Surveying.

AN OUTLINE OF THE METHOD OF CONDUCTING A TRIGONOMETRICAL SURVEY, for the Formation of Geographical and Topographical Maps and Plans, Military Reconnaissance, Levelling, &c., with Useful Problems, Formulæ, and Tables. By Lieut.-General FROME, R.E. Fourth Edition, Revised and partly Re-written by Major General Sir CHARLES WARREN, G.C.M.G., R.E. With 19 Plates and 115 Woodcuts. Royal 8vo, 16s. cloth.

"The simple fact that a fourth edition has been called for is the best testimony to its merits No words of praise from us can strengthen the position so well and so steadily maintained by this work. Sir Charles Warren has revised the entire work, and made such additions as were necessary to bring every portion of the contents up to the present date."—*Broad Arrow*.

Curves, Tables for Setting-out.

TABLES OF TANGENTIAL ANGLES AND MULTIPLES for Setting-out Curves from 5 to 200 Radius. By ALEXANDER BEAZELEY, M.Inst.C.E. Fourth Edition. Printed on 48 Cards, and sold in a cloth box, waistcoat-pocket size, 3s. 6d.

"Each table is printed on a small card, which, being placed on the theodolite, leaves the hand free to manipulate the instrument—no small advantage as regards the rapidity of work."—*Engineer*.

"Very handy; a man may know that all his day's work must fall on two of these cards, which he puts into his own card-case, and leaves the rest behind."—*Athenæum*.

Earthwork.

HANDY GENERAL EARTHWORK TABLES. Giving the Contents in Cubic Yards of Centre and Slopes of Cuttings and Embankments from 3 inches to 80 feet in Depth or Height, for use with either 66 feet Chain or 100 feet Chain. By J. H. WATSON BUCK, M.Inst.C.E. On a Sheet mounted in cloth case, 3s. 6d. [*Just published*.

Earthwork.

EARTHWORK TABLES. Showing the Contents in Cubic Yards of Embankments, Cuttings, &c., of Heights or Depths up to an average of 80 feet. By JOSEPH BROADBENT, C.E., and FRANCIS CAMPIN, C.E. Crown 8vo, 5s. cloth.

"The way in which accuracy is attained, by a simple division of each cross section into three elements, two in which are constant and one variable, is ingenious."—*Athenæum*.

Earthwork, Measurement of.

A MANUAL ON EARTHWORK. By ALEX. J. S. GRAHAM, C.E. With numerous Diagrams. Second Edition. 18mo, 2s. 6d. cloth.

"A great amount of practical information, very admirably arranged, and available for rough estimates, as well as for the more exact calculations required in the engineer's and contractor's offices."—*Artizan*.

Tunnelling.

PRACTICAL TUNNELLING. Explaining in detail the Setting-out of the works, Shaft-sinking and Heading-driving, Ranging the Lines and Levelling underground, Sub-Excavating, Timbering, and the Construction of the Brickwork of Tunnels, with the amount of Labour required for, and the Cost of, the various portions of the work. By FREDERICK W. SIMMS, M.Inst. C.E. Fourth Edition, Revised and Further Extended, including the Most Recent (1895) Examples of Sub-aqueous and other Tunnels, by D. KINNEAR CLARK, M.Inst. C.E. Imperial 8vo, with 34 Folding Plates and other Illustrations, £2 2s. cloth. [*Just published*.

"The estimation in which Mr. Simms's book on tunnelling has been held for over thirty years cannot be more truly expressed than in the words of the late Prof. Rankine:—'The best source of information or the subject of tunnels is Mr. F. W. Simms's work on Practical Tunnelling.'"—*Architect*.

"It has been regarded from the first as a text-book of the subject. . . . Mr. Clark has added immensely to the value of the book."—*Engineer*.

Tunnel Shafts.

THE CONSTRUCTION OF LARGE TUNNEL SHAFTS: A Practical and Theoretical Essay. By J. H. WATSON BUCK, M.Inst.C.E., Resident Engineer, London and North-Western Railway. Illustrated with Folding Plates. Royal 8vo, 12s. cloth.

"Many of the methods given are of extreme practical value to the mason; and the observations on the form of arch, the rules for ordering the stone, and the construction of the templates will be found of considerable use. We commend the book to the engineering profession."—*Building News*.

"Will be regarded by civil engineers as of the utmost value, and calculated to save much time and obviate many mistakes."—*Colliery Guardian*.

Oblique Bridges.

A PRACTICAL AND THEORETICAL ESSAY ON OBLIQUE BRIDGES. With 13 large Plates. By the late GEORGE WATSON BUCK, M.I.C.E. Fourth Edition, revised by his Son, J. H. WATSON BUCK, M.I.C.E. and with the addition of Description to Diagrams for Facilitating the Construction of Oblique Bridges, by W. H. BARLOW, M.I.C.E. Roy. 8vo, 12s. cl.

"The standard text-book for all engineers regarding skew arches is Mr. Buck's treatise, and it would be impossible to consult a better."—*Engineer.*

"Mr. Buck's treatise is recognised as a standard text-book, and his treatment has divested the subject of many of the intricacies supposed to belong to it. As a guide to the engineer and architect, on a confessedly difficult subject, Mr. Buck's work is unsurpassed."—*Building News.*

Cast and Wrought Iron Bridge Construction.

A COMPLETE AND PRACTICAL TREATISE ON CAST AND WROUGHT IRON BRIDGE CONSTRUCTION, *including Iron Foundations.* In Three Parts—Theoretical, Practical, and Descriptive. By WILLIAM HUMBER, A.M.Inst.C.E., and M.Inst.M.E. Third Edition, Revised and much improved, with 115 Double Plates (20 of which now first appear in this edition), and numerous Additions to the Text. In Two Vols., imp. 4to, £6 16s. 6d. half-bound in morocco.

"A very valuable contribution to the standard literature of civil engineering. In addition to elevations, plans and sections, large scale details are given which very much enhance the instructive worth of those illustrations."—*Civil Engineer and Architect's Journal.*

Iron Bridges.

IRON BRIDGES OF MODERATE SPAN: Their Construction and Erection. By HAMILTON WELDON PENDRED, late Inspector of Ironwork to the Salford Corporation. With 40 Illustrations. 12mo, 2s. cloth.

"Students and engineers should obtain this book for constant and practical use."—*Colliery Guardian.*

Oblique Arches.

A PRACTICAL TREATISE ON THE CONSTRUCTION OF OBLIQUE ARCHES. By JOHN HART. Third Edition, with Plates. Imperial 8vo, 8s. cloth.

Statics, Graphic and Analytic.

GRAPHIC AND ANALYTIC STATICS, *in their Practical Application to the Treatment of Stresses in Roofs, Solid Girders, Lattice, Bowstring and Suspension Bridges, Braced Iron Arches and Piers, and other Frameworks.* By R. HUDSON GRAHAM, C.E. Containing Diagrams and Plates to Scale. With numerous Examples, many taken from existing Structures. Specially arranged for Class-work in Colleges and Universities. Second Edition, Revised and Enlarged. 8vo, 16s. cloth.

"Mr. Graham's book will find a place wherever graphic and analytic statics are used or studied."—*Engineer.*

"The work is excellent from a practical point of view, and has evidently been prepared with much care. The directions for working are ample, and are illustrated by an abundance of well-selected examples. It is an excellent text-book for the practical draughtsman."—*Athenæum.*

Girders, Strength of.

GRAPHIC TABLE FOR FACILITATING THE COMPUTATION OF THE WEIGHTS OF WROUGHT IRON AND STEEL GIRDERS, etc., for Parliamentary and other Estimates. By J. H. WATSON BUCK, M.Inst.C.E. On a Sheet, 2s. 6d.

Strains, Calculation of.

A HANDY BOOK FOR THE CALCULATION OF STRAINS IN GIRDERS AND SIMILAR STRUCTURES, AND THEIR STRENGTH. Consisting of Formulæ and Corresponding Diagrams, with numerous details for Practical Application, &c. By WILLIAM HUMBER, A-M.Inst.C.E., &c. Fifth Edition. Crown 8vo, nearly 100 Woodcuts and 3 Plates, 7s. 6d. cloth

"The formulæ are neatly expressed, and the diagrams good."—*Athenæum.*

"We heartily commend this really *handy* book to our engineer and architect readers."—*English Mechanic.*

Trusses.

TRUSSES OF WOOD AND IRON. *Practical Applications of Science in Determining the Stresses, Breaking Weights, Safe Loads, Scantlings, and Details of Construction,* with Complete Working Drawings. By WILLIAM GRIFFITHS, Surveyor, Assistant Master, Tranmere School of Science and Art. Oblong 8vo, 4s. 6d. cloth.

"This handy little book enters so minutely into every detail connected with the construction of roof trusses, that no student need be ignorant of these matters."—*Practical Engineer.*

Strains in Ironwork.

THE STRAINS ON STRUCTURES OF IRONWORK; with Practical Remarks on Iron Construction. By F. W. SHEILDS, M.Inst.C.E, Second Edition, with 5 Plates. Royal 8vo, 5s. cloth.

"The student cannot find a better little book on this subject."—*Engineer.*

Barlow's Strength of Materials, enlarged by Humber.

A TREATISE ON THE STRENGTH OF MATERIALS; with Rules for Application in Architecture, the Construction of Suspension Bridges, Railways, &c. By PETER BARLOW, F.R.S. A New Edition, Revised by his Sons, P. W. BARLOW, F.R.S., and W. H. BARLOW, F.R.S.; to which are added, Experiments by HODGKINSON, FAIRBAIRN, and KIRKALDY; and Formulæ for Calculating Girders, &c. Arranged and Edited by WM. HUMBER, A-M.Inst.C.E. Demy 8vo, 400 pp., with 19 large Plates and numerous Woodcuts, 18s. cloth.

"Valuable alike to the student, tyro, and the experienced practitioner, it will always rank in future, as it has hitherto done, as the standard treatise on that particular subject."—*Engineer.*

"There is no greater authority than Barlow."—*Building News.*

"As a scientific work of the first class, it deserves a foremost place on the bookshelves of every civil engineer and practical mechanic."—*English Mechanic.*

Cast Iron and other Metals, Strength of.

A PRACTICAL ESSAY ON THE STRENGTH OF CAST IRON AND OTHER METALS. By THOMAS TREDGOLD, C.E. Fifth Edition, including HODGKINSON'S Experimental Researches. 8vo, 12s. cloth.

Railway Working.

SAFE RAILWAY WORKING. *A Treatise on Railway Accidents: Their Cause and Prevention; with a Description of Modern Appliances and Systems.* By CLEMENT E. STRETTON, C.E. With Illustrations and Coloured Plates. Third Edition, Enlarged. Crown 8vo, 3s. 6d.

"A book for the engineer, the directors, the managers; and, in short, all who wish for information on railway matters will find a perfect encyclopædia in 'Safe Railway Working.'"—*Railway Review.*

"We commend the remarks on railway signalling to all railway managers, especially where a uniform code and practice is advocated."—*Herepath's Railway Journal.*

"The author may be congratulated on having collected, in a very convenient form, much valuable information on the principal questions affecting the safe working of railways."—*Railway Engineer.*

Heat, Expansion by.

EXPANSION OF STRUCTURES BY HEAT. By JOHN KEILY, C.E., late of the Indian Public Works and Victorian Railway Departments. Crown 8vo, 3s. 6d. cloth.

SUMMARY OF CONTENTS.

Section I. FORMULAS AND DATA.
Section II. METAL BARS.
Section III. SIMPLE FRAMES.
Section IV. COMPLEX FRAMES AND PLATES.
Section V. THERMAL CONDUCTIVITY.
Section VI. MECHANICAL FORCE OF HEAT.
Section VII. WORK OF EXPANSION AND CONTRACTION.
Section VIII. SUSPENSION BRIDGES.
Section IX. MASONRY STRUCTURES.

"The aim the author has set before him, viz., to show the effects of heat upon metallic and other structures, is a laudable one, for this is a branch of physics upon which the engineer or architect can find but little reliable and comprehensive data in books."—*Builder.*

"Whoever is concerned to know the effect of changes of temperature on such structures as suspension bridges and the like, could not do better than consult Mr. Keily's valuable and handy exposition of the geometrical principles involved in these changes."—*Scotsman.*

Field Fortification.

A TREATISE ON FIELD FORTIFICATION, THE ATTACK OF FORTRESSES, MILITARY MINING, AND RECONNOITRING. By Colonel I. S. MACAULAY, late Professor of Fortification in the R.M.A., Woolwich. Sixth Edition. Crown 8vo, with separate Atlas of 12 Plates, 12s. cloth.

MR. HUMBER'S GREAT WORK ON MODERN ENGINEERING.

Complete in Four Volumes, imperial 4to, price £12 12s., half-morocco. Each Volume sold separately as follows:—

A RECORD OF THE PROGRESS OF MODERN ENGINEER-ING. FIRST SERIES. Comprising Civil, Mechanical, Marine, Hydraulic, Railway, Bridge, and other Engineering Works, &c. By WILLIAM HUMBER, A-M.Inst.C.E., &c. Imp. 4to, with 36 Double Plates, drawn to a large scale, Photographic Portrait of John Hawkshaw, C.E., F.R.S., &c., and copious descriptive Letterpress, Specifications, &c., £3 3s. half-morocco.

List of the Plates and Diagrams.

Victoria Station and Roof, L. B. & S. C. R. (8 plates); Southport Pier (2 plates); Victoria Station and Roof, L. C. & D. and G. W. R. (6 plates); Roof of Cremorne Music Hall; Bridge over G. N. Railway; Roof of Station, Dutch Rhenish Rail (2 plates); Bridge over the Thames, West London Extension Railway (5 plates); Armour Plates; Suspension Bridge, Thames (4 plates); The Allen Engine; Suspension Bridge, Avon (3 plates); Underground Railway (3 plates).

"Handsomely lithographed and printed. It will find favour with many who desire to preserve in a permanent form copies of the plans and specifications prepared for the guidance of the contractors for many important engineering works."—*Engineer.*

HUMBER'S PROGRESS OF MODERN ENGINEERING. SECOND SERIES. Imp. 4to, with 36 Double Plates, Photographic Portrait of Robert Stephenson, C.E., M.P., F.R.S., &c., and copious descriptive Letterpress, Specifications, &c., £3 3s. half-morocco.

List of the Plates and Diagrams.

Birkenhead Docks, Low Water Basin (15 plates); Charing Cross Station Roof, C. C. Railway (3 plates); Digswell Viaduct, Great Northern Railway; Robbery Wood Viaduct, Great Northern Railway; Iron Permanent Way; Clydach Viaduct, Merthyr, Tredegar, and Abergavenny Railway; Ebbw Viaduct, Merthyr, Tredegar, and Abergavenny Railway; College Wood Viaduct, Cornwall Railway; Dublin Winter Palace Roof (3 plates); Bridge over the Thames, L. C. & D. Railway (6 plates); Albert Harbour, Greenock (4 plates).

"Mr. Humber has done the profession good and true service, by the fine collection of examples he has here brought before the profession and the public."—*Practical Mechanic's Journal.*

HUMBER'S PROGRESS OF MODERN ENGINEERING. THIRD SERIES. Imp. 4to, with 40 Double Plates, Photographic Portrait of J. R. M'Clean, late Pres. Inst. C.E., and copious descriptive Letterpress, Specifications, &c., £3 3s. half-morocco.

List of the Plates and Diagrams.

MAIN DRAINAGE, METROPOLIS.—*North Side.*—Map showing Interception of Sewers; Middle Level Sewer (2 plates); Outfall Sewer, Bridge over River Lea (3 plates); Outfall Sewer, Bridge over Marsh Lane, North Woolwich Railway, and Bow and Barking Railway Junction; Outfall Sewer, Bridge over Bow and Barking Railway (3 plates); Outfall Sewer, Bridge over East London Waterworks' Feeder (2 plates); Outfall Sewer, Reservoir (2 plates); Outfall Sewer, Tumbling Bay and Outlet; Outfall Sewer, Penstocks. *South Side.*—Outfall Sewer, Bermondsey Branch (2 plates); Outfall Sewer, Reservoir and Outlet (4 plates); Outfall Sewer, Filth Hoist; Sections of Sewers (North and South Sides).
THAMES EMBANKMENT.—Section of River Wall; Steamboat Pier, Westminster (2 plates); Landing Stairs between Charing Cross and Waterloo Bridges; York Gate (2 plates); Overflow and Outlet at Savoy Street Sewer (3 plates); Steamboat Pier, Waterloo Bridge (3 plates); Junction of Sewers, Plans and Sections; Gullies, Plans and Sections; Rolling Stock Granite and Iron Forts.

"The drawings have a constantly increasing value, and whoever desires to possess clear representations of the two great works carried out by our Metropolitan Board will obtain Mr. Humber's volume."—*Engineer.*

HUMBER'S PROGRESS OF MODERN ENGINEERING. FOURTH SERIES. Imp. 4to, with 36 Double Plates, Photographic Portrait of John Fowler, late Pres. Inst. C.E., and copious descriptive Letterpress, Specifications, &c., £3 3s. half-morocco.

List of the Plates and Diagrams.

Abbey Mills Pumping Station, Main Drainage, Metropolis (4 plates); Barrow Docks (5 plates); Manquis Viaduct, Santiago and Valparaiso Railway (2 plates); Adam's Locomotive, St. Helen's Canal Railway (2 plates); Cannon Street Station Roof, Charing Cross Railway (3 plates); Road Bridge over the River Moka (2 plates); Telegraphic Apparatus for Mesopotamia; Viaduct over the River Wye, Midland Railway (3 plates); St. Germans Viaduct, Cornwall Railway (2 plates); Wrought-Iron Cylinder for Diving Bell; Millwall Docks (6 plates); Milroy's Patent Excavator; Metropolitan District Railway (6 plates); Harbours, Ports, and Breakwaters (3 plates).

"We gladly welcome another year's issue of this valuable publication from the able pen of Mr. Humber. The accuracy and general excellence of this work are well known, while its usefulness in giving the measurements and details of some of the latest examples of engineering, as carried out by the most eminent men in the profession, cannot be too highly prized."—*Artisan.*

16   CROSBY LOCKWOOD & SON'S CATALOGUE.

THE POPULAR WORKS OF MICHAEL REYNOLDS
("THE ENGINE DRIVER'S FRIEND").

Locomotive-Engine Driving.

LOCOMOTIVE-ENGINE DRIVING: *A Practical Manual for Engineers in charge of Locomotive Engines.* By MICHAEL REYNOLDS, Member of the Society of Engineers, formerly Locomotive Inspector L. B. and S. C. R. Ninth Edition. Including a KEY TO THE LOCOMOTIVE ENGINE. With Illustrations and Portrait of Author. Crown 8vo, 4s. 6d. cloth.

"Mr. Reynolds has supplied a want, and has supplied it well. We can confidently recommend the book, not only to the practical driver, but to everyone who takes an interest in the performance of locomotive engines."—*The Engineer.*

"Mr. Reynolds has opened a new chapter in the literature of the day. Of the practical utility of this admirable treatise, we have to speak in terms of warm commendation."—*Athenæum.*

"Evidently the work of one who knows his subject thoroughly."—*Railway Service Gazette.*

"Were the cautions and rules given in the book to become part of the every-day working of our engine-drivers, we might have fewer distressing accidents to deplore."—*Scotsman.*

Stationary Engine Driving.

STATIONARY ENGINE DRIVING: *A Practical Manual for Engineers in charge of Stationary Engines.* By MICHAEL REYNOLDS. Fifth Edition, Enlarged. With Plates and Woodcuts. Crown 8vo, 4s. 6d. cloth.

"The author is thoroughly acquainted with his subjects, and his advice on the various points treated is clear and practical. . . . He has produced a manual which is an exceedingly useful one for the class for whom it is specially intended."—*Engineering.*

"Our author leaves no stone unturned. He is determined that his readers shall not only know something about the stationary engine, but all about it."—*Engineer.*

"An engineman who has mastered the contents of Mr. Reynolds's book will require but little actual experience with boilers and engines before he can be trusted to look after them."—*English Mechanic.*

The Engineer, Fireman, and Engine-Boy.

THE MODEL LOCOMOTIVE ENGINEER, FIREMAN, and ENGINE-BOY. Comprising a Historical Notice of the Pioneer Locomotive Engines and their Inventors. By MICHAEL REYNOLDS. Second Edition, with Revised Appendix. With numerous Illustrations and Portrait of George Stephenson. Crown 8vo, 4s. 6d. cloth. [*Just published.*

"From the technical knowledge of the author it will appeal to the railway man of to-day more forcibly than anything written by Dr. Smiles. . . . The volume contains information of a technical kind, and facts that every driver should be familiar with."—*English Mechanic.*

"We should be glad to see this book in the possession of everyone in the kingdom who has ever laid, or is to lay, hands on a locomotive engine."—*Iron.*

Continuous Railway Brakes.

CONTINUOUS RAILWAY BRAKES: *A Practical Treatise on the several Systems in Use in the United Kingdom; their Construction and Performance.* With copious Illustrations and numerous Tables. By MICHAEL REYNOLDS. Large crown 8vo, 9s. cloth.

"A popular explanation of the different brakes. It will be of great assistance in forming public opinion, and will be studied with benefit by those who take an interest in the brake."—*English Mechanic.*

"Written with sufficient technical detail to enable the principle and relative connection of the various parts of each particular brake to be readily grasped."—*Mechanical World.*

Engine-Driving Life.

ENGINE-DRIVING LIFE: *Stirring Adventures and Incidents in the Lives of Locomotive-Engine Drivers.* By MICHAEL REYNOLDS. Third and Cheaper Edition. Crown 8vo, 1s. 6d. cloth.

"From first to last perfectly fascinating. Wilkie Collins's most thrilling conceptions are thrown into the shade by true incidents, endless in their variety, related in every page."—*North British Mail.*

"Anyone who wishes to get a real insight into railway life cannot do better than read 'Engine-Driving Life' for himself; and if he once take it up he will find that the author's enthusiasm and real love of the engine-driving profession will carry him on till he has read every page."—*Saturday Review.*

Pocket Companion for Enginemen.

THE ENGINEMAN'S POCKET COMPANION AND PRACTICAL EDUCATOR FOR ENGINEMEN, BOILER ATTENDANTS, AND MECHANICS. By MICHAEL REYNOLDS. With Forty-five Illustrations and numerous Diagrams. Third Edition, Revised. Royal 18mo, 3s. 6d., strongly bound for pocket wear.

"This admirable work is well suited to accomplish its object, being the honest workmanship of a competent engineer."—*Glasgow Herald.*

"A most meritorious work, giving in a succinct and practical form all the information an engineminder desirous of mastering the scientific principles of his daily calling would require."—*The Miller.*

"A boon to those who are striving to become efficient mechanics."—*Daily Chronicle.*

MARINE ENGINEERING, SHIPBUILDING, NAVIGATION, etc.

Pocket-Book for Naval Architects and Shipbuilders,
THE NAVAL ARCHITECT'S AND SHIPBUILDER'S POCKET-BOOK of Formulæ, Rules, and Tables, and MARINE ENGINEER'S AND SURVEYOR'S Handy Book of Reference. By CLEMENT MACKROW, Member of the Institution of Naval Architects, Naval Draughtsman. Sixth Edition, Revised. 700 pages, with upwards of 300 Illustrations. Fcap., 12s 6d. strongly bound in leather. [*Just published.*

SUMMARY OF CONTENTS.

SIGNS AND SYMBOLS, DECIMAL FRACTIONS.— TRIGONOMETRY. — PRACTICAL GEOMETRY. — MENSURATION. — CENTRES AND MOMENTS OF FIGURES.— MOMENTS OF INERTIA AND RADII OF GYRATION. — ALGEBRAICAL EXPRESSIONS FOR SIMPSON'S RULES.— MECHANICAL PRINCIPLES. — CENTRE OF GRAVITY.—LAWS OF MOTION.—DISPLACEMENT, CENTRE OF BUOYANCY.— CENTRE OF GRAVITY OF SHIP'S HULL. —STABILITY CURVES AND METACENTRES.—SEA AND SHALLOW-WATER WAVES.—ROLLING OF SHIPS.—PROPULSION AND RESISTANCE OF VESSELS. —SPEED TRIALS.—SAILING, CENTRE OF EFFORT.—DISTANCES DOWN RIVERS, COAST LINES.—STEERING AND RUDDERS OF VESSELS.—LAUNCHING CALCULATIONS AND VELOCITIES.—WEIGHT OF MATERIAL AND GEAR.—GUN PARTICULARS AND WEIGHT.—STANDARD GAUGES.—RIVETED JOINTS AND RIVETING.—STRENGTH AND TESTS OF MATERIALS. — BINDING AND SHEARING STRESSES, ETC.—STRENGTH OF SHAFTING, PILLARS, WHEELS, ETC. — HYDRAULIC DATA, ETC.—CONIC SECTIONS, CATENARIAN CURVES.—MECHANICAL POWERS, WORK. — BOARD OF TRADE REGULATIONS FOR BOILERS AND ENGINES. — BOARD OF TRADE REGULATIONS FOR SHIPS.—LLOYD'S RULES FOR BOILERS.—LLOYD'S WEIGHT OF CHAINS.—LLOYD'S SCANTLINGS FOR SHIPS.—DATA OF ENGINES AND VESSELS.- SHIPS' FITTINGS AND TESTS.— SEASONING PRESERVING TIMBER.— MEASUREMENT OF TIMBER.—ALLOYS, PAINTS, VARNISHES. — DATA FOR STOWAGE. — ADMIRALTY TRANSPORT REGULATIONS. — RULES FOR HORSEPOWER, SCREW PROPELLERS, ETC.— PERCENTAGES FOR BUTT STRAPS, ETC. —PARTICULARS OF YACHTS.—MASTING AND RIGGING VESSELS.—DISTANCES OF FOREIGN PORTS. — TONNAGE TABLES. — VOCABULARY OF FRENCH AND ENGLISH TERMS. — ENGLISH WEIGHTS AND MEASURES.—FOREIGN WEIGHTS AND MEASURES.—DECIMAL EQUIVALENTS. — FOREIGN MONEY.— DISCOUNT AND WAGE TABLES. — USEFUL NUMBERS AND READY RECKONERS —TABLES OF CIRCULAR MEASURES.— TABLES OF AREAS OF AND CIRCUMFERENCES OF CIRCLES.—TABLES OF AREAS OF SEGMENTS OF CIRCLES.— TABLES OF SQUARES AND CUBES AND ROOTS OF NUMBERS. — TABLES OF LOGARITHMS OF NUMBERS.—TABLES OF HYPERBOLIC LOGARITHMS.—TABLES OF NATURAL SINES, TANGENTS, ETC.— TABLES OF LOGARITHMIC SINES, TANGENTS, ETC.

"In these days of advanced knowledge a work like this is of the greatest value. It contains a vast amount of information. We unhesitatingly say that it is the most valuable compilation for its specific purpose that has ever been printed. No naval architect, engineer, surveyor, or seaman, wood or iron shipbuilder, can afford to be without this work."—*Nautical Magazine.*

"Should be used by all who are engaged in the construction or designs of vessels. . . . Will be found to contain the most useful tables and formulæ required by shipbuilders, carefully collected from the best authorities, and put together in a popular and simple form."—*Engineer.*

"The professional shipbuilder has now, in a convenient and accessible form, reliable data for solving many of the numerous problems that present themselves in the course of his work."—*Iron.*

"There is no doubt that a pocket-book of this description must be a necessity in the shipbuilding trade. . . . The volume contains a mass of useful information clearly expressed and presented in a handy form."—*Marine Engineer.*

Marine Engineering.

MARINE ENGINES AND STEAM VESSELS (A Treatise on). By ROBERT MURRAY, C.E. Eighth Edition, thoroughly Revised, with considerable Additions by the Author and by GEORGE CARLISLE, C.E., Senior Surveyor to the Board of Trade at Liverpool. 12mo, 5s. cloth boards.

"Well adapted to give the young steamship engineer or marine engine and boiler maker a general introduction into his practical work."—*Mechanical World*

"We feel sure that this thoroughly revised edition will continue to be as popular in the future as it has been in the past, as, for its size, it contains more useful information than any similar treatise."—*Industries.*

"As a compendious and useful guide to engineers of our mercantile and royal naval services, we should say it cannot be surpassed."—*Building News.*

"The information given is both sound and sensible, and well qualified to direct young seagoing hands on the straight road to the extra chief's certificate. . . . Most useful to surveyors, inspectors, draughtsmen, and young engineers."—*Glasgow Herald.*

English-French Dictionary of Sea Terms.

TECHNICAL DICTIONARY OF SEA TERMS, PHRASES AND WORDS USED IN THE ENGLISH & FRENCH LANGUAGES. (English-French, French-English). For the Use of Seamen, Engineers, Pilots, Ship-builders, Ship-owners and Ship-brokers. Compiled by W. PIRRIE, late of the African Steamship Company. Fcap. 8vo, 5s. cloth limp.

[*Just published.*

"This volume will be highly appreciated by seamen, engineers, pilots, shipbuilders and ship-owners. It will be found wonderfully accurate and complete."—*Scotsman.*

"A very useful dictionary, which has long been wanted by French and English engineers, masters, officers and others."—*Shipping World.*

Pocket-Book for Marine Engineers.

A POCKET-BOOK OF USEFUL TABLES AND FORMULÆ FOR MARINE ENGINEERS. By FRANK PROCTOR, A.I.N.A. Third Edition. Royal 32mo, leather, gilt edges, with strap, 4s.

"We recommend it to our readers as going far to supply a long-felt want."—*Naval Science.*

"A most useful companion to all marine engineers."—*United Service Gazette.*

Introduction to Marine Engineering.

ELEMENTARY ENGINEERING: *A Manual for Young Marine Engineers and Apprentices.* In the Form of Questions and Answers on Metals, Alloys, Strength of Materials, Construction and Management of Marine Engines and Boilers, Geometry, &c. &c. With an Appendix of Useful Tables. By JOHN SHERREN BREWER, Government Marine Surveyor, Hong-kong. Third Edition. Small crown 8vo, 1s. 6d. cloth.

"Contains much valuable information for the class for whom it is intended, especially in the chapters on the management of boilers and engines."—*Nautical Magazine.*

"A useful introduction to the more elaborate text-books."—*Scotsman.*

"To a student who has the requisite desire and resolve to attain a thorough knowledge, Mr. Brewer offers decidedly useful help."—*Athenæum.*

Navigation.

PRACTICAL NAVIGATION. Consisting of THE SAILOR'S SEA-BOOK, by JAMES GREENWOOD and W. H. ROSSER; together with the requisite Mathematical and Nautical Tables for the Working of the Problems, by HENRY LAW, C.E., and Professor J. R. YOUNG. Illustrated. 12mo, 7s. strongly half-bound.

Sailmaking.

THE ART AND SCIENCE OF SAILMAKING. By SAMUEL B. SADLER, Practical Sailmaker, late in the employment of Messrs. Ratsey and Lapthorne, of Cowes and Gosport. With Plates and other Illustrations. Small 4to, 12s. 6d. cloth.

"This work is very ably written, and is illustrated by diagrams and carefully-worked calculations. The work should be in the hands of every sailmaker, whether employer or employed, as it cannot fail to assist them in the pursuit of their important avocations."—*Isle of Wight Herald.*

"This extremely practical work gives a complete education in all the branches of the manufacture cutting out, roping, seaming, and goring. It is copiously illustrated, and will form a first-rate text-book and guide."—*Portsmouth Times.*

"The author of this work has rendered a distinct service to all interested in the art of sailmaking. The subject of which he treats is a congenial one. Mr. Sadler is a practical sailmaker and has devoted years of careful observation and study to the subject; and the results of the experience thus gained he has set forth in the volume before us."—*Steamship.*

Chain Cables.

CHAIN CABLES AND CHAINS. Comprising Sizes and Curves of Links, Studs, &c., Iron for Cables and Chains, Chain Cable and Chain Making, Forming and Welding Links, Strength of Cables and Chains, Certificates for Cables, Marking Cables, Prices of Chain Cables and Chains, Historical Notes, Acts of Parliament, Statutory Tests, Charges for Testing, List of Manufacturers of Cables, &c. &c. By THOMAS W. TRAILL, F.E.R.N., M. Inst. C.E., Engineer Surveyor in Chief, Board of Trade, Inspector of Chain Cable and Anchor Proving Establishments, and General Superintendent, Lloyd's Committee on Proving Establishments. With numerous Tables, Illustrations and Lithographic Drawings. Folio, £2 2s. cloth.

"It contains a vast amount of valuable information. Nothing seems to be wanting to make it a complete and standard work of reference on the subject."—*Nautical Magazine.*

MINING AND METALLURGY.

Mining Machinery.

MACHINERY FOR METALLIFEROUS MINES: A Practical Treatise for Mining Engineers, Metallurgists, and Managers of Mines. By E. HENRY DAVIES, M.E., F.G.S. Crown 8vo, 580 pp., with upwards of 300 Illustrations, 12s. 6d. cloth. [*Just published.*

"Mr. Davies, in this handsome volume, has done the advanced student and the manager of mines good service. Almost every kind of machinery in actual use is carefully described, and the woodcuts and plates are good."—*Athenæum.*

"From cover to cover the work exhibits all the same characteristics which excite the confidence and attract the attention of the student as he peruses the first page. The work may safely be recommended. By its publication the literature connected with the industry will be enriched, and the reputation of its author enhanced."—*Mining Journal.*

"Mr. Davies has endeavoured to bring before his readers the best of everything in modern mining appliances. His work carries internal evidence of the author's impartiality, and this constitutes one of the great merits of the book. Throughout his work the criticisms are based on his own or other reliable experience.'—*Iron and Steel Trades' Journal.*

"The work deals with nearly every class of machinery or apparatus likely to be met with or required in connection with metalliferous mining, and is one which we have every confidence in recommending."—*Practical Engineer.*

Metalliferous Minerals and Mining.

A TREATISE ON METALLIFEROUS MINERALS AND MINING. By D. C. DAVIES, F.G.S., Mining Engineer, &c., Author of "A Treatise on Slate and Slate Quarrying." Fifth Edition, thoroughly Revised and much Enlarged, by his Son, E. HENRY DAVIES, M.E., F.G.S. With about 150 Illustrations. Crown 8vo, 12s. 6d. cloth.

"Neither the practical miner nor the general reader interested in mines can have a better book for his companion and his guide."—*Mining Journal.* [*Mining World.*

"We are doing our readers a service in calling their attention to this valuable work."

"A book that will not only be useful to the geologist, the practical miner, and the metallurgist but also very interesting to the general public."—*Iron.*

"As a history of the present state of mining throughout the world this book has a real value and it supplies an actual want."—*Athenæum.*

Earthy Minerals and Mining.

A TREATISE ON EARTHY & OTHER MINERALS AND MINING. By D. C. DAVIES, F.G.S., Author of "Metalliferous Minerals," &c. Third Edition, revised and Enlarged, by his Son, E. HENRY DAVIES M.E., F.G.S. With about 100 Illustrations. Crown 8vo, 12s. 6d. cloth.

"We do not remember to have met with any English work on mining matters that contains the same amount of information packed in equally convenient form."—*Academy.*

"We should be inclined to rank it as among the very best of the handy technical and trades manuals which have recently appeared."—*British Quarterly Review.*

Metalliferous Mining in the United Kingdom.

BRITISH MINING: A Treatise on the History, Discovery, Practical Development, and Future Prospects of Metalliferous Mines in the United Kingdom. By ROBERT HUNT, F.R.S., Editor of "Ure's Dictionary of Arts, Manufactures, and Mines," &c. Upwards of 950 pp., with 230 Illustrations. Second Edition. Revised. Super-royal 8vo, £2 2s. cloth.

"One of the most valuable works of reference of modern times. Mr. Hunt, as Keeper of Mining Records of the United Kingdom, has had opportunities for such a task not enjoyed by anyone else and has evidently made the most of them. . . . The language and style adopted are good, and the treatment of the various subjects laborious, conscientious, and scientific."—*Engineering.*

"The book is, in fact, a treasure-house of statistical information on mining subjects, and we know of no other work embodying so great a mass of matter of this kind. Were this the only merit of Mr. Hunt's volume, it would be sufficient to render it indispensable in the library of everyone interested in the development of the mining and metallurgical industries of this country. —*Athenæum.*

"A mass of information not elsewhere available, and of the greatest value to those who may be interested in our great mineral industries."—*Engineer.*

Underground Pumping Machinery.

MINE DRAINAGE. Being a Complete and Practical Treatise on Direct-Acting Underground Steam Pumping Machinery, with a Description of a large number of the best known Engines, their General Utility and the Special Sphere of their Action, the Mode of their Application, and their merits compared with other forms of Pumping Machinery. By STEPHEN MICHELL. 8vo, 15s. cloth.

"Will be highly esteemed by colliery owners and lessees, mining engineers, and students generally who require to be acquainted with the best means of securing the drainage of mines. It is a most valuable work, and stands almost alone in the literature of steam pumping machinery."—*Colliery Guardian.*

'Much valuable information is given, so that the book is thoroughly worthy of an extensive circulation amongst practical men and purchasers of machinery."—*Mining Journal.*

Prospecting for Gold and other Metals.

THE PROSPECTOR'S HANDBOOK: A Guide for the Prospector and Traveller in Search of Metal-Bearing or other Valuable Minerals. By J. W. ANDERSON, M.A. (Camb.), F.R.G.S., Author of "Fiji and New Caledonia." Sixth Edition, thoroughly Revised and much Enlarged. Small crown 8vo, 3s. 6d. cloth ; or, 4s. 6d. leather, pocket-book form, with tuck.
[*Just published.*
"Will supply a much felt want, especially among Colonists, in whose way are so often thrown many mineralogical specimens the value of which it is difficult to determine."—*Engineer.*
"How to find commercial minerals, and how to identify them when they are found, are the leading points to which attention is directed. The author has managed to pack as much practical detail into his pages as would supply material for a book three times its size."—*Mining Journal.*

Mining Notes and Formulæ.

NOTES AND FORMULÆ FOR MINING STUDENTS. By JOHN HERMAN MERIVALE, M.A., Certificated Colliery Manager, Professor of Mining in the Durham College of Science, Newcastle-upon-Tyne. Third Edition, Revised and Enlarged. Small crown 8vo, 2s. 6d. cloth.
"Invaluable to anyone who is working up for an examination on mining subjects."—*Iron and Coal Trades Review.*
"The author has done his work in an exceedingly creditable manner, and has produced a book that will be of service to students, and those who are practically engaged in mining operations."—*Engineer.*

Handybook for Miners.

THE MINER'S HANDBOOK : A Handy Book of Reference on the Subjects of Mineral Deposits, Mining Operations, Ore Dressing, &c. For the Use of Students and others interested in Mining matters. Compiled by JOHN MILNE, F.R.S., Professor of Mining in the Imperial University of Japan. Revised Edition. Fcap. 8vo, 7s. 6d. leather. [*Just published.*
"Professor Milne's handbook is sure to be received with favour by all connected with mining, and will be extremely popular among students."—*Athenæum.*

Miners' and Metallurgists' Pocket-Book.

A POCKET-BOOK FOR MINERS AND METALLURGISTS. Comprising Rules, Formulæ, Tables, and Notes, for Use in Field and Office Work. By F. DANVERS POWER, F.G.S., M.E. Fcap. 8vo, 9s. leather.
"This excellent book is an admirable example of its kind, and ought to find a large sale amongst English-speaking prospectors and mining engineers."—*Engineering.*
"A useful *vade-mecum* containing a mass of rules, formulæ, tables, and various other information, necessary for daily eference."—*Iron.*

Mineral Surveying and Valuing.

THE MINERAL SURVEYOR AND VALUER'S COMPLETE GUIDE, *comprising a Treatise on Improved Mining Surveying and the Valuation of Mining Properties, with New Traverse Tables.* By WM. LINTERN. Third Edition, Enlarged. 12mo, 4s. cloth.
"A valuable and thoroughly trustworthy guide."—*Iron and Coal Trades Review.*

Asbestos and its Uses.

ASBESTOS: *Its Properties, Occurrence, and Uses.* With some Account of the Mines of Italy and Canada. By ROBERT H. JONES. With Eight Collotype Plates and other Illustrations. Crown 8vo, 12s. 6d. cloth.
"An interesting and invaluable work."—*Colliery Guardian.*

Iron, Metallurgy of.

METALLURGY OF IRON. Containing History of Iron Manufacture, Methods of Assay, and Analyses of Iron Ores, Processes of Manufacture of Iron and Steel, &c. By H. BAUERMAN, F.G.S., A.R.S.M. With numerous Illustrations. Sixth Edition, Enlarged. 12mo, 5s. 6d. cloth.
"Carefully written, it has the merit of brevity and conciseness, as to less important points; while all material matters are very fully and thoroughly entered into."—*Standard.*

Slate Quarrying, &c.

SLATE AND SLATE QUARRYING, Scientific, Practical and Commercial. By D. C. DAVIES, F.G.S., Mining Engineer, &c. With numerous Illustrations and Folding Plates. Third Edition, 12mo, 3s. cloth.
"One of the best and best-balanced treatises on a special subject that we have met with."—*Engineer.*

Colliery Management.

THE COLLIERY MANAGER'S HANDBOOK: A Comprehensive Treatise on the Laying-out and Working of Collieries, Designed as a Book of Reference for Colliery Managers, and for the Use of Coal-Mining Students preparing for First-class Certificates. By CALEB PAMELY, Mining Engineer and Surveyor; Member of the North of England Institute of Mining and Mechanical Engineers; and Member of the South Wales Institute of Mining Engineers. With nearly 700 Plans, Diagrams, and other Illustrations. Third Edition, Revised and Enlarged. Medium 8vo, about 900 pages. Price £1 5s. strongly bound. *[Just published.*

SUMMARY OF CONTENTS.

GEOLOGY. — SEARCH FOR COAL. — MINERAL LEASES AND OTHER HOLDINGS. — SHAFT SINKING. — FITTING UP THE SHAFT AND SURFACE ARRANGEMENTS. — STEAM BOILERS AND THEIR FITTINGS. — TIMBERING AND WALLING. — NARROW WORK AND METHODS OF WORKING. — UNDERGROUND CONVEYANCE. — DRAINAGE. — THE GASES MET WITH IN MINES; VENTILATION. — ON THE FRICTION OF AIR IN MINES. — THE PRIESTMAN OIL ENGINE; PETROLEUM AND NATURAL GAS — SURVEYING AND PLANNING. — LIGHTING; SAFETY LAMPS AND FIRE-DAMP DETECTORS. — SUNDRY AND INCIDENTAL OPERATIONS AND APPLIANCES. — COLLIERY EXPLOSIONS. — MISCELLANEOUS QUESTIONS AND ANSWERS.

Appendix: SUMMARY OF REPORT OF H.M. COMMISSIONERS ON ACCIDENTS IN MINES.

"There can be no doubt that it is the best book on coal-mining."—J. T. ROBSON, Esq., *H.M.'s Inspector of Mines, South Wales District.*
"Mr. Pamely's work is eminently suited to the purpose for which it is intended—being clear, interesting, exhaustive, rich in detail, and up to date, giving descriptions of the very latest machines in every department. . . . A mining engineer could scarcely go wrong who followed this work."—*Colliery Guardian.*
"This is the most complete 'all round' work on coal-mining published in the English language. . . . No library of coal-mining books is complete without it."—*Colliery Engineer* (Scranton, Pa., U.S.A.).
"Mr. Pamely's work is in all respects worthy of our admiration. No person in any responsible position connected with mines should be without a copy."—*Westminster Review.*

Coal and Iron.

THE COAL AND IRON INDUSTRIES OF THE UNITED KINGDOM. Comprising a Description of the Coal Fields, and of the Principal Seams of Coal, with Returns of their Produce and its Distribution, and Analyses of Special Varieties. Also an Account of the occurrence of Iron Ores in Veins or Seams; Analyses of each Variety; and a History of the Rise and Progress of Pig Iron Manufacture. By RICHARD MEADE, Assistant Keeper of Mining Records. With Maps. 8vo, £1 8s. cloth.

"The book is one which must find a place on the shelves of all interested in coal and iron production, and in the iron, steel, and other metallurgical industries."—*Engineer.*
"Of this book we may unreservedly say that it is the best of its class which we have ever met. . . . A book of reference which no one engaged in the iron or coal trades should omit from his library."—*Iron and Coal Trades Review.*

Coal Mining.

COAL AND COAL MINING: *A Rudimentary Treatise on.* By the late Sir WARINGTON W. SMYTH, M.A., F.R.S., Chief Inspector of the Mines of the Crown. Seventh Edition, Revised and Enlarged. With numerous Illustrations. 12mo, 4s. cloth boards.

"As an outline is given of every known coal-field in this and other countries, as well as of the principal methods of working, the book will doubtless interest a very large number of readers."—*Mining Journal.*

Subterraneous Surveying.

SUBTERRANEOUS SURVEYING, Elementary and Practical Treatise on, with and without the Magnetic Needle. By THOMAS FENWICK, Surveyor of Mines, and THOMAS BAKER, C.E. Illust. 12mo, 3s. cloth boards.

Granite Quarrying.

GRANITES AND OUR GRANITE INDUSTRIES. By GEORGE F. HARRIS, F.G.S., Membre de la Société Belge de Géologie, Lecturer on Economic Geology at the Birkbeck Institution, &c. With Illustrations. Crown 8vo, 2s. 6d. cloth.

"A clearly and well-written manual on the granite industry."—*Scotsman.*
"An interesting work, which will be deservedly esteemed."—*Colliery Guardian.*
"An exceedingly interesting and valuable monograph on a subject which has hitherto received unaccountably little attention in the shape of systematic literary treatment."—*Scottish Leader.*

Gold, Metallurgy of.

THE METALLURGY OF GOLD: A Practical Treatise on the Metallurgical Treatment of Gold-bearing Ores. Including the Processes of Concentration, Chlorination and Extraction by Cyanide, and the Assaying, Melting, and Refining of Gold. By M. EISSLER, Mining Engineer and Metallurgical Chemist, formerly Assistant Assayer of the U. S. Mint, San Francisco. Fourth Edition, Enlarged. With about 250 Illustrations and numerous Folding Plates and Working Drawings. 8vo, 16s. cloth   [*Just published.*

"This book thoroughly deserves its title of a 'Practical Treatise.' The whole process of gold milling, from the breaking of the quartz to the assay of the bullion, is described in clear and orderly narrative and with much, but not too much, fulness of detail."—*Saturday Review.*
"The work is a storehouse of information and valuable data, and we strongly recommend it to all professional men engaged in the gold-mining industry."—*Mining Journal.*

Gold Extraction.

THE CYANIDE PROCESS OF GOLD EXTRACTION: and its Practical Application on the Witwatersrand Gold Fields in South Africa. By M. EISSLER, M.E., Mem. Inst. Mining and Metallurgy, Author of "The Metallurgy of Gold," &c. With Diagrams and Working Drawings. Large crown 8vo, 7s. 6d. cloth.   [*Just published.*

"This book is just what was needed to acquaint mining men with the actual working of a process which is not only the most popular, but is, as a general rule, the most successful for the extraction of gold from tailings."—*Mining Journal.*
"The work will prove invaluable to all interested in gold mining, whether metallurgists or as investors."—*Chemical News.*

Silver, Metallurgy of.

THE METALLURGY OF SILVER: A Practical Treatise on the Amalgamation, Roasting, and Lixiviation of Silver Ores. Including the Assaying, Melting and Refining, of Silver Bullion. By M. EISSLER, Author of "The Metallurgy of Gold," &c. Third Edition. With 150 Illustrations. Crown 8vo, 10s. 6d. cloth.   [*Just published.*

"A practical treatise, and a technical work which we are convinced will supply a long-felt want amongst practical men, and at the same time be of value to students and others indirectly connected with the industries."—*Mining Journal.*
"From first to last the book is thoroughly sound and reliable."—*Colliery Guardian.*
"For chemists, practical miners, assayers, and investors alike, we do not know of any work on the subject so handy and yet so comprehensive."—*Glasgow Herald.*

Lead, Metallurgy of.

THE METALLURGY OF ARGENTIFEROUS LEAD: A Practical Treatise on the Smelting of Silver-Lead Ores and the Refining of Lead Bullion. Including Reports on various Smelting Establishments and Descriptions of Modern Smelting Furnaces and Plants in Europe and America. By M. EISSLER, M.E., Author of "The Metallurgy of Gold," &c. Crown 8vo, 400 pp., with 183 Illustrations, 12s. 6d. cloth.

"The numerous metallurgical processes, which are fully and extensively treated of, embrace all the stages experienced in the passage of the lead from the various natural states to its issue from the refinery as an article of commerce."—*Practical Engineer.*
"The present volume fully maintains the reputation of the author. Those who wish to obtain a thorough insight into the present state of this industry cannot do better than read this volume, and all mining engineers cannot fail to find many useful hints and suggestions in it."—*Industries.*
"It is most carefully written and illustrated with capital drawings and diagrams. In fact, it is the work of an expert for experts, by whom it will be prized as an indispensable text-book."—*Bristol Mercury.*

Iron Mining.

THE IRON ORES OF GREAT BRITAIN AND IRELAND: Their Mode of Occurrence, Age, and Origin, and the Methods of Searching for and Working them, with a Notice of some of the Iron Ores of Spain. By J. D. KENDALL, F.G.S., Mining Engineer. Crown 8vo, 16s. cloth.

"The author has a thorough practical knowledge of his subject, and has supplemented a careful study of the available literature by unpublished information derived from his own observations. The result is a very useful volume which cannot fail to be of value to all interested in the iron industry of the country."—*Industries.*
"Mr. Kendall is a great authority on this subject and writes from personal observation."—*Colliery Guardian.*
"Mr. Kendall's book is thoroughly well done. In it there are the outlines of the history of ore mining in every centre and there is everything that we want to know as to the character of the ores of each district, their commercial value and the cost of working them."—*Iron and Steel Trades Journal.*

ELECTRICITY, ELECTRICAL ENGINEERING, etc.

Dynamo Management.

THE MANAGEMENT OF DYNAMOS: A Handybook of Theory and Practice for the Use of Mechanics, Engineers, Students and others in Charge of Dynamos. By G. W. LUMMIS PATERSON. With numerous Illustrations. Crown 8vo, 3s. 6d. cloth. [*Just published*.

"An example which deserves to be taken as a model by other authors. The subject is treated in a manner which any intelligent man who is fit to be entrusted with charge of an engine should be able to understand. It is a useful book to all who make, tend or employ electric machinery."—*Architect*.

"A most satisfactory book from a practical point of view. We strongly commend it to the attention of every electrical engineering student."—*Daily Chronicle*.

Electrical Engineering.

THE ELECTRICAL ENGINEER'S POCKET-BOOK OF MODERN RULES, FORMULÆ, TABLES, AND DATA. By H. R. KEMPE, M.Inst.E.E., A.M.Inst.C.E., Technical Officer, Postal Telegraphs, Author of "A Handbook of Electrical Testing," &c. Second Edition, thoroughly Revised, with Additions. Royal 32mo, oblong, 5s. leather.

"There is very little in the shape of formulæ or data which the electrician is likely to want in a hurry which cannot be found in its pages."—*Practical Engineer*.

"A very useful book of reference for daily use in practical electrical engineering and its various applications to the industries of the present day."—*Iron*.

"It is the best book of its kind."—*Electrical Engineer*.

"Well arranged and compact. The 'Electrical Engineer's Pocket-Book' is a good one."—*Electrician*. [*Review*.

"Strongly recommended to those engaged in the various electrical industries."—*Electrical*

Electric Lighting.

ELECTRIC LIGHT FITTING: A Handbook for Working Electrical Engineers, embodying Practical Notes on Installation Management. By JOHN W. URQUHART, Electrician, Author of "Electric Light," &c. With numerous Illustrations. Second Edition, Revised, with Additional Chapters. Crown 8vo, 5s. cloth.

"This volume deals with what may be termed the mechanics of electric lighting, and is addressed to men who are already engaged in the work or are training for it. The work traverses a great deal of ground, and may be read as a sequel to the same author's useful work on 'Electric Light.'"—*Electrician*.

"The book is well worth the perusal of the workmen for whom it is written."—*Electrical Review*.

"We have read this book with a good deal of pleasure. We believe that the book will be of use to practical workmen, who will not be alarmed by finding mathematical formulæ which they are unable to understand."—*Electrical Plant*.

Electric Light.

ELECTRIC LIGHT: Its Production and Use. Embodying Plain Directions for the Treatment of Dynamo-Electric Machines, Batteries, Accumulators, and Electric Lamps. By J. W. URQUHART, C.E., Author of "Electric Light Fitting," "Electroplating," &c. Fifth Edition, carefully Revised, with Large Additions and 145 Illustrations. Crown 8vo, 7s. 6d. cloth.

"The whole ground of electric lighting is more or less covered and explained in a very clear and concise manner."—*Electrical Review*.

"Contains a good deal of very interesting information, especially in the parts where the author gives dimensions and working costs."—*Electrical Engineer*.

"A miniature *vade-mecum* of the salient facts connected with the science of electric lighting."—*Electrician*.

"You cannot have a better book than 'Electric Light,' by Urquhart."—*Engineer*.

"The book is by far the best that we have yet met with on the subject."—*Athenæum*.

Construction of Dynamos.

DYNAMO CONSTRUCTION: A Practical Handbook for the Use of Engineer Constructors and Electricians-in-Charge. Embracing Framework Building, Field Magnet and Armature Winding and Grouping, Compounding, &c. With Examples of leading English, American, and Continental Dynamos and Motors. By J. W. URQUHART, Author of "Electric Light," "Electric Light Fitting," &c. Second Edition, Revised and Enlarged. With 114 Illustrations. Crown 8vo, 7s. 6d. cloth. [*Just published*.

"Mr. Urquhart's book is the first one which deals with these matters in such a way that the engineering student can understand them. The book is very readable, and the author leads his readers up to difficult subjects by reasonably simple tests."—*Engineering Review*.

"The author deals with his subject in a style so popular as to make his volume a handbook of great practical value to engineer constructors and electricians in charge."—*Scotsman*.

"'Dynamo Construction' more than sustains the high character of the author's previous publications. It is sure to be widely read by the large and rapidly-increasing number of practical electricians."—*Glasgow Herald*.

CROSBY LOCKWOOD & SON'S CATALOGUE.

New Dictionary of Electricity.

THE STANDARD ELECTRICAL DICTIONARY. A Popular Dictionary of Words and Terms Used in the Practice of Electrical Engineering. Containing upwards of 3,000 Definitions. By T. O'CONNOR SLOANE, A.M., Ph.D., Author of "The Arithmetic of Electricity," &c. Crown 8vo, 630 pp., 350 Illustrations, 7s. 6d. cloth. [*Just published.*

"The work has many attractive features in it, and is beyond doubt, a well put together and useful publication. The amount of ground covered may be gathered from the fact that in the index about 5,000 references will be found. The inclusion of such comparatively modern words as 'impedence,' 'reluctance,' &c., shows that the author has desired to be up to date, and indeed there are other indications of carefulness of compilation. The work is one which does the author great credit and it should prove of great value, especially to students."—*Electrical Review.*

"Very complete and contains a large amount of useful information."—*Industries.*

"An encyclopædia of electrical science in the compass of a dictionary. The information given is sound and clear. The book is well printed, well illustrated, and well up to date, and may be confidently recommended."—*Builder.*

"The volume is excellently printed and illustrated, and should form part of the library of every one who is connected with electrical matters."—*Hardware Trade Journal.*

Electric Lighting of Ships.

ELECTRIC SHIP-LIGHTING: A Handbook on the Practical Fitting and Running of Ship's Electrical Plant. For the Use of Shipowners and Builders, Marine Electricians, and Sea-going Engineers-in-Charge. By J. W. URQUHART, C.E. With 88 Illustrations. Crown 8vo, 7s. 6d. cloth.

"The subject of ship electric lighting is one of vast importance in these days, and Mr. Urquhart is to be highly complimented for placing such a valuable work at the service of the practical marine electrician."—*The Steamship.*

"Distinctly a book which of its kind stands almost alone, and for which there should be a demand."—*Electrical Review.*

Country House Electric Lighting.

ELECTRIC LIGHT FOR COUNTRY HOUSES: A Practical Handbook on the Erection and Running of Small Installations, with particulars of the Cost of Plant and Working. By J. H. KNIGHT. Crown 8vo, 1s. wrapper. [*Just published.*

Electric Lighting.

THE ELEMENTARY PRINCIPLES OF ELECTRIC LIGHTING. By ALAN A. CAMPBELL SWINTON, Associate I.E.E. Third Edition, Enlarged and Revised. With 16 Illustrations. Crown 8vo, 1s. 6d. cloth.

"Anyone who desires a short and thoroughly clear exposition of the elementary principles of electric-lighting cannot do better than read this little work."—*Bradford Observer.*

Dynamic Electricity.

THE ELEMENTS OF DYNAMIC ELECTRICITY AND MAGNETISM. By PHILIP ATKINSON, A.M., Ph.D., Author of "The Elements of Electric Lighting," &c. Cr. 8vo, with 120 Illustrations, 10s. 6d. cl.

Electric Motors, &c.

THE ELECTRIC TRANSFORMATION OF POWER and its Application by the Electric Motor, including Electric Railway Construction. By P. ATKINSON, A.M., Ph.D., Author of "The Elements of Electric Lighting," &c. With 96 Illustrations. Crown 8vo, 7s. 6d. cloth.

Dynamo Construction.

HOW TO MAKE A DYNAMO: *A Practical Treatise for Amateurs.* Containing numerous Illustrations and Detailed Instructions for Constructing a Small Dynamo, to Produce the Electric Light. By ALFRED CROFTS. Fifth Edition, Revised and Enlarged. Crown 8vo, 2s. cloth.

"The instructions given in this unpretentious little book are sufficiently clear and explicit to enable any amateur mechanic possessed of average skill and the usual tools to be found in an amateur's workshop, to build a practical dynamo machine."—*Electrician.*

Text Book of Electricity.

THE STUDENT'S TEXT-BOOK OF ELECTRICITY. By HENRY M. NOAD, F.R.S. 630 pages, with 470 Illustrations. Cheaper Edition. Crown 8vo, 9s. cloth. [*Just published.*

Electricity.

A MANUAL OF ELECTRICITY: Including Galvanism, Magnetism, Dia-Magnetism, Electro-Dynamics. By HENRY M. NOAD, Ph D., F.R.S. Fourth Edition (1859). 8vo, £1 4s. cloth.

ARCHITECTURE, BUILDING, etc.

Building Construction.
PRACTICAL BUILDING CONSTRUCTION: A Handbook for Students Preparing for Examinations, and a Book of Reference for Persons Engaged in Building. By JOHN P. ALLEN, Surveyor, Lecturer on Building Construction at the Durham College of Science, Newcastle. Medium 8vo, 450 pages, with 1,000 Illustrations. 12s. 6d. cloth. [*Just published*.

"This volume is one of the most complete expositions of building construction we have seen. It contains all that is necessary to prepare students for the various examinations in building construction."—*Building News*.

"The author depends nearly as much on his diagrams as on his type. The pages suggest the hand of a man of experience in building operations—and the volume must be a blessing to many teachers as well as to students.'—*The Architect*.

"The work is sure to prove a formidable rival to great and small competitors alike, and bids fair to take a permanent place as a favourite students' text-book. The large number of illustrations deserve particular mention for the great merit they possess for purposes of reference, in exactly corresponding to convenient scales."—*Jour. Inst. Brit. Archts.*

Masonry.
PRACTICAL MASONRY: A Guide to the Art of Stone Cutting. Comprising the Construction, Setting-Out, and Working of Stairs, Circular Work, Arches, Niches, Domes, Pendentives, Vaults, Tracery Windows, &c. For the Use of Students, Masons and other Workmen. By WILLIAM R. PURCHASE, Building Inspector to the Town of Hove. Royal 8vo, 134 pages, including 50 Lithographic Plates (about 400 separate Diagrams), 7s. 6d. cloth. [*Just published*.

"The illustrations are well thought out and clear. The volume places within reach of the professional mason many useful data for solving the problems which present themselves day by day.'—*Glasgow Herald*.

The New Builder's Price Book, 1896.
LOCKWOOD'S BUILDER'S PRICE BOOK FOR 1896. A Comprehensive Handbook of the Latest Prices and Data for Builders, Architects, Engineers, and Contractors. By FRANCIS T. W. MILLER. 800 closely-printed pages, crown 8vo, 4s. cloth.

"This book is a very useful one, and should find a place in every English office connected with the building and engineering professions."—*Industries*.

"An excellent book of reference."—*Architect*.

"In its new and revised form this Price Book is what a work of this kind should be—comprehensive, reliable, well arranged, legible, and well bound."—*British Architect*.

New London Building Act, 1894.
THE LONDON BUILDING ACT, 1894; with the By-Laws and Regulations of the London County Council, and Introduction, Notes, Cases and Index. By ALEX. J. DAVID, B.A., LL.M. of the Inner Temple, Barrister-at-Law. Crown 8vo, 3s. 6d. cloth. [*Just published*.

"To all architects and district surveyors and builders, Mr. David's manual will be welcome."—*Building News*.

"The volume will doubtless be eagerly consulted by the building fraternity."—*Illustrated Carpenter and Builder*.

Concrete.
CONCRETE: ITS NATURE AND USES. A Book for Architects, Builders, Contractors, and Clerks of Works. By GEORGE L. SUTCLIFFE, A.R.I.B.A. Crown 8vo, 7s. 6d. cloth. [*Just published*.

"The author treats a difficult subject in a lucid manner. The manual fills a long-felt gap. It careful and exhaustive; equally useful as a student's guide and a architect's book of reference."—*Journal of Royal Institution of British Architects*.

"There is room for this new book, which will probably be for some time the standard work on the subject for a builder's purpose."—*Glasgow Herald*.

"A thoroughly useful and comprehensive work."—*British Architect*.

Mechanics for Architects.
THE MECHANICS OF ARCHITECTURE: A Treatise on Applied Mechanics, especially Adapted to the Use of Architects. By E. W. TARN, M.A., Author of "The Science of Building," &c. Second Edition, Enlarged. Illust. with 125 Diagrams. Cr. 8vo, 7s. 6d. cloth. [*Just published*.

"The book is a very useful and helpful manual of architectural mechanics, and really contains sufficient to enable a careful and painstaking student to grasp the principles bearing upon the majority of building problems. . . . Mr. Tarn has added, by this volume, to the debt of gratitude which is owing to him by architectural students for the many valuable works which he has produced for their use."—*The Builder*.

"The mechanics in the volume are really mechanics, and are harmoniously wrought in with the distinctive professional manner proper to the subject. —*The Schoolmaster*.

Designing Buildings.
THE DESIGN OF BUILDINGS: Being Elementary Notes on the Planning, Sanitation and Ornamentive Formation of Structures, based on Modern Practice. Illustrated with Nine Folding Plates. By W. WOODLEY, Assistant Master, Metropolitan Drawing Classes, &c. 8vo, 6s. cloth.

Sir Wm. Chambers's Treatise on Civil Architecture.
THE DECORATIVE PART OF CIVIL ARCHITECTURE. By Sir WILLIAM CHAMBERS, F.R.S. With Portrait, Illustrations, Notes, and an Examination of Grecian Architecture, by JOSEPH GWILT, F.S.A. Revised and Edited by W. H. LEEDS. 66 Plates, 4to, 21s. cloth.

Villa Architecture.
A HANDY BOOK OF VILLA ARCHITECTURE: Being a Series of Designs for Villa Residences in various Styles. With Outline Specifications and Estimates. By C. WICKES, Architect, Author of "The Spires and Towers of England," &c. 61 Plates, 4to, £1 11s. 6d. half-morocco.
"The whole of the designs bear evidence of their being the work of an artistic architect, and they will prove very valuable and suggestive."—*Building News.*

Text-Book for Architects.
THE ARCHITECT'S GUIDE: Being a Text-Book of Useful Information for Architects, Engineers, Surveyors, Contractors, Clerks of Works, &c. By F. ROGERS. Third Edition. Crown 8vo, 3s. 6d. cloth.
"As a text-book of useful information for architects, engineers, surveyors, &c., it would be hard to find a handier or more complete little volume."—*Standard.*

Linear Perspective.
ARCHITECTURAL PERSPECTIVE: The whole Course and Operations of the Draughtsman in Drawing a Large House in Linear Perspective. Illustrated by 43 Folding Plates. By F. O. FERGUSON. Second Edition, Enlarged. 8vo, 3s. 6d. boards. [*Just published.*
"It is the most intelligible of the treatises on this ill-treated subject that I have met with."—E. INGRESS BELL, Esq., in the *R.I.B.A. Journal.*

Architectural Drawing.
PRACTICAL RULES ON DRAWING, *for the Operative Builder and Young Student in Architecture.* By G. PYNE. 14 Plates, 4to, 7s. 6d., bds.

Vitruvius' Architecture.
THE ARCHITECTURE of MARCUS VITRUVIUS POLLIO. Translated by JOSEPH GWILT, F.S.A., F.R.A.S. New Edition, Revised by the Translator. With 23 Plates. Fcap. 8vo, 5s. cloth.

Designing, Measuring, and Valuing.
THE STUDENT'S GUIDE to the PRACTICE of MEASURING AND VALUING ARTIFICERS' WORK. Containing Directions for taking Dimensions, Abstracting the same, and bringing the Quantities into Bill, with Tables of Constants for Valuation of Labour, and for the Calculation of Areas and Solidities. Originally edited by EDWARD DOBSON, Architect. With Additions by E. WYNDHAM TARN, M.A. Sixth Edition. With 8 Plates and 63 Woodcuts. Crown 8vo, 7s. 6d. cloth.
"This edition will be found the most complete treatise on the principles of measuring and valuing artificers' work that has yet been published."—*Building News.*

Pocket Estimator and Technical Guide.
THE POCKET TECHNICAL GUIDE, MEASURER, AND ESTIMATOR FOR BUILDERS AND SURVEYORS. Containing Technical Directions for Measuring Work in all the Building Trades, Complete Specifications for Houses, Roads, and Drains, and an easy Method of Estimating the parts of a Building collectively. By A. C. BEATON. Seventh Edit. Waistcoat-pocket size, 1s. 6d. leather, gilt edges.
"No builder, architect, surveyor, or valuer should be without his 'Beaton.'"—*Building News.*

Donaldson on Specifications.
THE HANDBOOK OF SPECIFICATIONS; or, Practical Guide to the Architect, Engineer, Surveyor, and Builder, in drawing up Specifications and Contracts for Works and Constructions. Illustrated by Precedents of Buildings actually executed by eminent Architects and Engineers. By Professor T. L. DONALDSON, P.R.I.B.A., &c. New Edition. 8vo. with upwards of 1,000 pages of Text, and 33 Plates. £1 11s. 6d. cloth.
"Valuable as a record, and more valuable still as a book of precedents. . . . Suffice it to say that Donaldson's 'Handbook of Specifications' must be bought by all architects."—*Builder.*

Bartholomew and Rogers' Specifications.

SPECIFICATIONS FOR PRACTICAL ARCHITECTURE. A Guide to the Architect, Engineer, Surveyor, and Builder. With an Essay on the Structure and Science of Modern Buildings. Upon the Basis of the Work by ALFRED BARTHOLOMEW, thoroughly Revised, Corrected, and greatly added to by FREDERICK ROGERS, Architect. Third Edition, Revised, with Additions. With numerous Illustrations. Medium 8vo, 15s. cloth.

"The collection of specifications prepared by Mr. Rogers on the basis of Bartholomew's work is too well known to need any recommendation from us. It is one of the books with which every young architect must be equipped."—*Architect.*

House Building and Repairing.

THE HOUSE-OWNER'S ESTIMATOR; or, What will it Cost to Build, Alter, or Repair? A Price Book for Unprofessional People, as well as the Architectural Surveyor and Builder. By J. D. SIMON. Edited by F. T. W. MILLER, A.R.I.B.A. Fourth Edition. Crown 8vo, 3s. 6d. cloth.

"In two years it will repay its cost a hundred times over."—*Field.*

Construction.

THE SCIENCE OF BUILDING: An Elementary Treatise on the Principles of Construction. By E. WYNDHAM TARN, M.A., Architect. Third Edition, Revised and Enlarged. With 59 Engravings. Fcap. 8vo, 4s. cl.

A very valuable book, which we strongly recommend to all students."—*Builder.*

Building; Civil and Ecclesiastical.

A BOOK ON BUILDING, Civil and Ecclesiastical, including Church Restoration; with the Theory of Domes and the Great Pyramid, &c. By Sir EDMUND BECKETT, Bart., LL.D., F.R.AS. Fcap. 8vo, 5s. cloth.

"A book which is always amusing and nearly always instructive."—*Times.*

House Building.

DWELLING HOUSES, THE ERECTION OF. Illustrated by a Perspective View, Plans, Elevations and Sections of a Pair of Semi-Detached Villas, with the Specification, Quantities and Estimates. By S. H. BROOKS, Architect. Seventh Edition, thoroughly Revised. 12mo, 2s. 6d. cloth. [*Just published.*

Sanitary Houses, etc.

THE SANITARY ARRANGEMENT OF DWELLING-HOUSES: A Handbook for Householders and Owners of Houses. By A. J. WALLIS-TAYLER, A.M. Inst. C.E. With numerous Illustrations. Crown 8vo, 2s. 6d. cloth. [*Just published.*

"This book will be largely read; it will be of considerable service to the public. It is well arranged, easily read, and for the most part devoid of technical terms."—*Lancet.*

Ventilation of Buildings.

VENTILATION. A Text Book to the Practice of the Art of Ventilating Buildings. By W. P. BUCHAN, R.P. 12mo, 4s. cloth.

"Contains a great amount of useful practical information, as thoroughly interesting as it is technically reliable."—*British Architect.*

The Art of Plumbing.

PLUMBING. A Text Book to the Practice of the Art or Craft of the Plumber. By WILLIAM PATON BUCHAN, R.P. Sixth Edition. 4s. cloth.

"A text-book which may be safely put in the hands of every young plumber."—*Builder.*

Geometry for the Architect, Engineer, etc.

PRACTICAL GEOMETRY, for the Architect, Engineer, and Mechanic. Giving Rules for the Delineation and Application of various Geometrical Lines, Figures and Curves. By E. W. TARN, M.A., Architect. 8vo, 9s. cloth.

"No book with the same objects in view has ever been published in which the clearness of the rules laid down and the illustrative diagrams have been so satisfactory."—*Scotsman.*

The Science of Geometry.

THE GEOMETRY OF COMPASSES; or, Problems Resolved by the mere Description of Circles, and the use of Coloured Diagrams and Symbols. By OLIVER BYRNE. Coloured Plates. Crown 8vo, 3s. 6d. cloth.

CARPENTRY, TIMBER, etc.

Tredgold's Carpentry, Revised & Enlarged by Tarn.
THE ELEMENTARY PRINCIPLES OF CARPENTRY. A Treatise on the Pressure and Equilibrium of Timber Framing, the Resistance of Timber, and the Construction of Floors, Arches, Bridges, Roofs, Uniting Iron and Stone with Timber, &c. To which is added an Essay on the Nature and Properties of Timber, &c., with Descriptions of the kinds of Wood used in Building; also numerous Tables of the Scantlings of Timber for different purposes, the Specific Gravities of Materials, &c. By THOMAS TREDGOLD, C.E. With an Appendix of Specimens of Various Roofs of Iron and Stone, Illustrated. Seventh Edition, thoroughly revised and considerably enlarged by E. WYNDHAM TARN, M.A., Author of "The Science of Building," &c. With 61 Plates, Portrait of the Author, and several Woodcuts. In One large Vol., 4to, price £1 5s. cloth.

"Ought to be in every architect's and every builder's library."—*Builder.*

"A work whose monumental excellence must commend it wherever skilful carpentry is concerned. The author's principles are rather confirmed than impaired by time. The additional plates are of great intrinsic value."—*Building News.*

Carpentry.
CARPENTRY AND JOINERY. The Elementary Principles of Carpentry. Chiefly composed from the Standard Work of THOMAS TREDGOLD, C.E. With Additions, and a TREATISE ON JOINERY by E. W. TARN, M.A. Fifth Edition, Revised and Extended. 12mo, 3s. 6d. cloth.

⁎ ATLAS of Thirty-five Plates to accompany and illustrate the foregoing book. With Descriptive Letterpress. 4to, 6s. cloth.

"These two volumes form a complete treasury of carpentry and joinery, and should be in the hands of every carpenter and joiner in the empire."—*Iron.*

Woodworking Machinery.
WOODWORKING MACHINERY: Its Rise, Progress, and Construction. With Hints on the Management of Saw Mills and the Economical Conversion of Timber. Illustrated with Examples of Recent Designs by leading English, French, and American Engineers. By M. POWIS BALE, A.M.Inst.C.E., M.I.M.E. Second Edition, Revised, with large Additions. Large crown 8vo, 440 pp., 9s. cloth. [*Just published.*

"Mr. Bale is evidently an expert on the subject and he has collected so much information that the book is all-sufficient for builders and others engaged in the conversion of timber."—*Architect.*

"The most comprehensive compendium of wood-working machinery we have seen. The author is a thorough master of his subject."—*Building News.*

Saw Mills.
SAW MILLS: Their Arrangement and Management, and the Economical Conversion of Timber. (A Companion Volume to "Woodworking Machinery.") By M. POWIS BALE. Crown 8vo, 10s. 6d. cloth.

"The *administration* of a large sawing establishment is discussed, and the subject examined from a financial standpoint. Hence the size, shape, order, and disposition of saw-mills and the like are gone into in detail, and the course of the timber is traced from its reception to its delivery in its converted state. We could not desire a more complete or practical treatise."—*Builder.*

Nicholson's Carpentry.
THE CARPENTER'S NEW GUIDE; or, Book of Lines for Carpenters; comprising all the Elementary Principles essential for acquiring a knowledge of Carpentry. Founded on the late PETER NICHOLSON's Standard Work. New Edition, Revised by A. ASHPITEL, F.S.A. With Practical Rules on Drawing, by G. PYNE. With 74 Plates, 4to, £1 1s. cloth.

Circular Work.
CIRCULAR WORK IN CARPENTRY AND JOINERY: A Practical Treatise on Circular Work of Single and Double Curvature. By GEORGE COLLINGS. With Diagrams. Second Edit. 12mo, 2s. 6d. cloth limp.

"An excellent example of what a book of this kind should be. Cheap in price, clear in definition and practical in the examples selected."—*Builder.*

Handrailing.
HANDRAILING COMPLETE IN EIGHT LESSONS. On the Square-Cut System. By J. S. GOLDTHORP, Teacher of Geometry and Building Construction at the Halifax Mechanic's Institute. With Eight Plates and over 150 Practical Exercises. 4to, 3s. 6d. cloth.

"Likely to be of considerable value to joiners and others who take a pride in good work we heartily commend it to teachers and students."—*Timber Trades Journal.*

Handrailing and Stairbuilding.

A PRACTICAL TREATISE ON HANDRAILING: Showing New and Simple Methods for Finding the Pitch of the Plank, Drawing the Moulds, Bevelling, Jointing-up, and Squaring the Wreath. By GEORGE COLLINGS. Second Edition, Revised and Enlarged, to which is added A TREATISE ON STAIRBUILDING. 12mo, 2s. 6d. cloth limp.

"Will be found of practical utility in the execution of this difficult branch of joinery."—*Builder.*
"Almost every difficult phase of this somewhat intricate branch of joinery is elucidated by the aid of plates and explanatory letterpress."—*Furniture Gazette.*

Timber Merchant's Companion.

THE TIMBER MERCHANT'S AND BUILDER'S COMPANION. Containing New and Copious Tables of the Reduced Weight and Measurement of Deals and Battens, of all sizes, from One to a Thousand Pieces, and the relative Price that each size bears per Lineal Foot to any given Price per Petersburg Standard Hundred; the Price per Cube Foot of Square Timber to any given Price per Load of 50 Feet; the proportionate Value of Deals and Battens by the Standard, to Square Timber by the Load of 50 Feet; the readiest mode of ascertaining the Price of Scantling per Lineal Foot of any size, to any given Figure per Cube Foot, &c. &c. By WILLIAM DOWSING. Fourth Edition, Revised and Corrected. Cr. 8vo, 3s. cl.

"Everything is as concise and clear as it can possibly be made. There can be no doubt that every timber merchant and builder ought to possess it."—*Hull Advertiser.*
"We are glad to see a fourth edition of these admirable tables, which for correctness and simplicity of arrangement leave nothing to be desired."—*Timber Trades Journal.*

Practical Timber Merchant.

THE PRACTICAL TIMBER MERCHANT. Being a Guide for the use of Building Contractors, Surveyors, Builders, &c., comprising useful Tables for all purposes connected with the Timber Trade, Marks of Wood, Essay on the Strength of Timber, Remarks on the Growth of Timber, &c. By W. RICHARDSON. Second Edition. Fcap. 8vo, 3s. 6d. cloth.

"This handy manual contains much valuable information for the use of timber merchants, builders, foresters, and all others connected with the growth, sale, and manufacture of timber."—*Journal of Forestry.*

Packing-Case Makers, Tables for.

PACKING-CASE TABLES; showing the number of Superficial Feet in Boxes or Packing-Cases, from six inches square and upwards. By W. RICHARDSON, Timber Broker. Third Edition. Oblong 4to, 3s. 6d. cl.

"Invaluable labour-saving tables."—*Ironmonger.*
"Will save much labour and calculation."—*Grocer.*

Superficial Measurement.

THE TRADESMAN'S GUIDE TO SUPERFICIAL MEASUREMENT. Tables calculated from 1 to 200 inches in length, by 1 to 108 inches in breadth. For the use of Architects, Surveyors, Engineers, Timber Merchants, Builders, &c. By JAMES HAWKINGS. Fourth Edition. Fcap., 3s. 6d. cloth.

"A useful collection of tables to facilitate rapid calculation of surfaces. The exact area of any surface of which the limits have been ascertained can be instantly determined. The book will be found of the greatest utility to all engaged in building operations."—*Scotsman.*
"These tables will be found of great assistance to all who require to make calculations in superficial measurement."—*English Mechanic.*

Forestry.

THE ELEMENTS OF FORESTRY. Designed to afford Information concerning the Planting and Care of Forest Trees for Ornament or Profit, with Suggestions upon the Creation and Care of Woodlands. By F. B. HOUGH. Large crown 8vo, 10s. cloth.

Timber Importer's Guide.

THE TIMBER IMPORTER'S, TIMBER MERCHANT'S, AND BUILDER'S STANDARD GUIDE. By RICHARD E. GRANDY. Comprising an Analysis of Deal Standards, Home and Foreign, with Comparative Values and Tabular Arrangements for fixing Net Landed Cost on Baltic and North American Deals, including all intermediate Expenses, Freight Insurance, &c. &c. Together with copious Information for the Retailer and Builder. Third Edition, Revised. 12mo, 2s. cloth limp.

"Everything it pretends to be: built up gradually, it leads one from a forest to a treenail and throws in, as a makeweight, a host of material concerning bricks, columns, cisterns, &c.'"—*English Mechanic.*

DECORATIVE ARTS, etc.

Woods and Marbles (Imitation of).

SCHOOL OF PAINTING FOR THE IMITATION OF WOODS AND MARBLES, as Taught and Practised by A. R. VAN DER BURG and P. VAN DER BURG, Directors of the Rotterdam Painting Institution. Royal folio, 18¾ by 12½ in., Illustrated with 24 full-size Coloured Plates; also 12 plain Plates, comprising 154 Figures. Second and Cheaper Edition. Price £1 11s. 6d.

List of Plates.

1. Various Tools required for Wood Painting—2, 3. Walnut: Preliminary Stages of Graining and Finished Specimen—4. Tools used for Marble Painting and Method of Manipulation—6. St. Remi Marble: Earlier Operations and Finished Specimen—7. Methods of Sketching different Grains, Knots, &c.—8, 9. Ash: Preliminary Stages and Finished Specimen—10. Methods of Sketching Marble Grains—11, 12. Breche Marble: Preliminary Stages of Working and Finished Specimen—13. Maple: Methods of Producing the different Grains—14, 15. Bird's-eye Maple: Preliminary Stages and Finished Specimen—16. Methods of Sketching the different Species of White Marble—17, 18. White Marble: Preliminary Stages of Process and Finished Specimen—19. Mahogany: Specimens of various Grains and Methods of Manipulation—20, 21. Mahogany: Earlier Stages and Finished Specimen—22, 23, 24. Sienna Marble: Varieties of Grain, Preliminary Stages and Finished Specimen—25, 26, 27. Juniper Wood: Methods of producing Grain, &c.; Preliminary Stages and Finished Specimen—28, 29, 30. Vert de Mer Marble: Varieties of Grain and Methods of Working Unfinished and Finished Specimens—31, 32, 33. Oak: Varieties of Grain, Tools Employed, and Methods of Manipulation, Preliminary Stages and Finished Specimen—34, 35, 36. Waulsort Marble: Varieties of Grain, Unfinished and Finished Specimens.

"Those who desire to attain skill in the art of painting woods and marbles will find advantage in consulting this book. . . . Some of the Working Men's Clubs should give their young men the opportunity to study it."—*Builder.*

"A comprehensive guide to the art. The explanations of the processes, the manipulation and management of the colours, and the beautifully executed plates will not be the least valuable to the student who aims at making his work a faithful transcript of nature."—*Building News.*

House Decoration.

ELEMENTARY DECORATION. A Guide to the Simpler Forms of Everyday Art. Together with PRACTICAL HOUSE DECORATION. By JAMES W. FACEY. With numerous Illustrations. In One Vol., 5s. strongly half-bound

House Painting, Graining, etc.

HOUSE PAINTING, GRAINING, MARBLING, AND SIGN WRITING, A Practical Manual of. By ELLIS A. DAVIDSON. Sixth Edition. With Coloured Plates and Wood Engravings. 12mo, 6s. cloth boards.

"A mass of information, of use to the amateur and of value to the practical man."—*English Mechanic.*

Decorators, Receipts for.

THE DECORATOR'S ASSISTANT: A Modern Guide to Decorative Artists and Amateurs, Painters, Writers, Gilders, &c. Containing upwards of 600 Receipts, Rules and Instructions; with a variety of Information for General Work connected with every Class of Interior and Exterior Decorations, &c. Sixth Edition. 152 pp., crown 8vo, 1s. in wrapper

"Full of receipts of value to decorators, painters, gilders, &c. The book contains the gist of larger treatises on colour and technical processes. It would be difficult to meet with a work so full of varied information on the painter's art."—*Building News.*

Moyr Smith on Interior Decoration.

ORNAMENTAL INTERIORS, ANCIENT AND MODERN. By J. MOYR SMITH. Super-royal 8vo, with 32 full-page Plates and numerous smaller Illustrations, handsomely bound in cloth, gilt top, price 18s.

"The book is well illustrated and handsomely got up, and contains some true criticism and a great many good examples of decorative treatment."—*The Builder.*

DECORATIVE ARTS, etc. 31

British and Foreign Marbles.

MARBLE DECORATION and the Terminology of British and Foreign Marbles. A Handbook for Students. By GEORGE H. BLAGROVE, Author of "Shoring and its Application," &c. With 28 Illustrations. Crown 8vo, 3s. 6d. cloth.

"This most useful and much wanted handbook should be in the hands of every architect and builder."—*Building World.*

"A carefully and usefully written treatise; the work is essentially practical."—*Scotsman.*

Marble Working, etc.

MARBLE AND MARBLE WORKERS: A Handbook for Architects, Artists, Masons, and Students. By ARTHUR LEE, Author of "A Visit to Carrara," "The Working of Marble," &c. Small crown 8vo, 2s. cloth.

"A really valuable addition to the technical literature of architects and masons."—*Building News.*

DELAMOTTE'S WORKS ON ILLUMINATION AND ALPHABETS.

A PRIMER OF THE ART OF ILLUMINATION, for the Use of Beginners: with a Rudimentary Treatise on the Art, Practical Directions for its Exercise, and Examples taken from Illuminated MSS., printed in Gold and Colours. By F. DELAMOTTE. New and Cheaper Edition. Small 4to, 6s. ornamental boards.

"The examples of ancient MSS. recommended to the student, which, with much good sense, the author chooses from collections accessible to all, are selected with judgment and knowledge, as well as taste."—*Athenæum.*

ORNAMENTAL ALPHABETS, Ancient and Mediæval, from the Eighth Century, with Numerals; including Gothic, Church-Text, large and small, German, Italian, Arabesque, Initials for Illumination, Monograms, Crosses, &c. &c., for the use of Architectural and Engineering Draughtsmen, Missal Painters, Masons, Decorative Painters, Lithographers, Engravers, Carvers, &c. &c. Collected and Engraved by F. DELAMOTTE, and printed in Colours. New and Cheaper Edition. Royal 8vo, oblong, 2s. 6d. ornamental boards.

"For those who insert enamelled sentences round gilded chalices, who blazon shop legends over shop-doors, who letter church walls with pithy sentences from the Decalogue, this book will be useful."—*Athenæum.*

EXAMPLES OF MODERN ALPHABETS, Plain and Ornamental; including German, Old English, Saxon, Italic, Perspective, Greek, Hebrew, Court Hand, Engrossing, Tuscan, Riband, Gothic, Rustic, and Arabesque; with several Original Designs, and an Analysis of the Roman and Old English Alphabets, large and small, and Numerals, for the use of Draughtsmen, Surveyors, Masons, Decorative Painters, Lithographers, Engravers, Carvers, &c. Collected and Engraved by F. DELAMOTTE, and printed in Colours. New and Cheaper Edition. Royal 8vo, oblong, 2s. 6d. ornamental boards.

"There is comprised in it every possible shape into which the letters of the alphabet and numerals can be formed, and the talent which has been expended in the conception of the various plain and ornamental letters is wonderful."—*Standard.*

MEDIÆVAL ALPHABETS AND INITIALS FOR ILLUMINATORS. By F. G. DELAMOTTE. Containing 21 Plates and Illuminated Title, printed in Gold and Colours. With an Introduction by J. WILLIS BROOKS. Fourth and Cheaper Edition. Small 4to, 4s. ornamental boards.

"A volume in which the letters of the alphabet come forth glorified in gilding and all the colours of the prism interwoven and intertwined and intermingled."—*Sun.*

THE EMBROIDERER'S BOOK OF DESIGN. Containing Initials, Emblems, Cyphers, Monograms, Ornamental Borders, Ecclesiastical Devices, Mediæval and Modern Alphabets, and National Emblems. Collected by F. DELAMOTTE, and printed in Colours. Oblong royal 8vo, 1s. 6d. ornamental wrapper.

"The book will be of great assistance to ladies and young children who are endowed with the art of plying the needle in this most ornamental and useful pretty work."—*East Anglian Times.*

Wood Carving.

INSTRUCTIONS IN WOOD-CARVING, for Amateurs; with Hints on Design. By A LADY. With Ten Plates. New and Cheaper Edition. Crown 8vo, 2s. in emblematic wrapper.

"The handicraft of the wood-carver, so well as a book can impart it, may be learnt from 'A Lady's' publication."—*Athenæum.*

NATURAL SCIENCE, etc.

The Heavens and their Origin.

THE VISIBLE UNIVERSE: Chapters on the Origin and Construction of the Heavens. By J. E. Gore, F.R.A.S. Illustrated by 6 Stellar Photographs and 12 Plates. 8vo, 16s. cloth.

"A valuable and lucid summary of recent astronomical theory, rendered more valuable and attractive by a series of stellar photographs and other illustrations."—*The Times.*

"In presenting a clear and concise account of the present state of our knowledge, Mr. Gore has made a valuable addition to the literature of the subject."—*Nature.*

"As interesting as a novel, and instructive withal; the text being made still more luminous by stellar photographs and other illustrations. . . . A most valuable book."—*Manchester Examiner.*

"One of the finest works on astronomical science that has recently appeared in our language.
Leeds Mercury.

The Constellations.

STAR GROUPS: A Student's Guide to the Constellations. By J. Ellard Gore, F.R.A.S., M.R.I.A., &c., Author of "The Visible Universe," "The Scenery of the Heavens." With 30 Maps. Small 4to, 5s. cloth, silvered.

"A knowledge of the principal constellations visible in our latitudes may be easily acquired from the thirty maps and accompanying text contained in this work."—*Nature.*

"The volume contains thirty maps showing stars of the sixth magnitude—the usual naked-eye limit—and each is accompanied by a brief commentary, adapted to facilitate recognition and bring to notice objects of special interest. For the purpose of a preliminary survey of the 'midnight pomp' of the heavens, nothing could be better than a set of delineations averaging scarcely twenty square inches in area, and including nothing that cannot at once be identified."—*Saturday Review.*

"A very compact and handy guide to the constellations."—*Athenæum.*

Astronomical Terms.

AN ASTRONOMICAL GLOSSARY; or, Dictionary of Terms used in Astronomy. With Tables of Data and Lists of Remarkable and Interesting Celestial Objects. By J. Ellard Gore, F.R.A.S., Author of "The Visible Universe," &c. Small crown 8vo, 2s. 6d. cloth.

"A very useful little work for beginners in astronomy, and not to be despised by more advanced students."—*The Times.*

"Astronomers of all kinds will be glad to have it for reference."—*Guardian.*

The Microscope.

THE MICROSCOPE: Its Construction and Management, including Technique, Photo-micrography, and the Past and Future of the Microscope. By Dr. Henri van Heurck. Re-Edited and Augmented from the Fourth French Edition, and Translated by Wynne E. Baxter, F.G.S, 400 pages, with upwards of 250 Woodcuts. Imp. 8vo, 18s. cloth.

"A translation of a well-known work, at once popular and comprehensive."—*Times.*

"The translation is as felicitous as it is accurate."—*Nature.*

The Microscope.

PHOTO-MICROGRAPHY. By Dr. H. van Heurck. Extracted from the above Work. Royal 8vo, with Illustrations, 1s. sewed.

Astronomy.

ASTRONOMY. By the late Rev. Robert Main, F.R.S. Third Edition, Revised, by Wm. T. Lynn, B.A., F.R.A.S., 12mo, 2s. cloth.

"A sound and simple treatise, and a capital book for beginners."—*Knowledge.*

Recent and Fossil Shells.

A MANUAL OF THE MOLLUSCA: Being a Treatise on Recent and Fossil Shells. By S. P. Woodward, A.L.S., F.G.S. With an Appendix on *Recent and Fossil Conchological Discoveries*, by Ralph Tate, A.L.S., F.G.S. With 23 Plates and upwards of 300 Woodcuts. Reprint of Fourth Edition, 1880. Crown 8vo, 7s. 6d. cloth.

"A most valuable storehouse of conchological and geological information."—*Science Gossip.*

Geology and Genesis.

THE TWIN RECORDS OF CREATION; or, Geology and Genesis: their Perfect Harmony and Wonderful Concord. By George W. Victor Le Vaux. Fcap. 8vo, 5s. cloth.

"A valuable contribution to the evidences of Revelation, and disposes very conclusively of the arguments of those who would set God's Works against God's Word. No real difficulty is shirked, and no sophistry is left unexposed. —*The Rock.*

Geology.

RUDIMENTARY TREATISE ON GEOLOGY, PHYSICAL AND HISTORICAL. With especial reference to the British series of Rocks. By R. Tate, F.G.S. With 250 Illustrations. 12mo, 5s. cloth boards.

NATURAL SCIENCE, etc. 33

DR. LARDNER'S COURSE OF NATURAL PHILOSOPHY.

HANDBOOK OF MECHANICS. Re-written and Enlarged by B. LOEWY, F.R.A.S. Post 8vo, 6s. cloth.
"Mr. Loewy has carefully revised the book, and brought it up to modern requirements."—*Nature.*

HANDBOOK OF HYDROSTATICS & PNEUMATICS. Enlarged by B. LOEWY, F.R.A.S. Post 8vo, 5s. cloth.
"For those 'who desire to attain an accurate knowledge of physical science without the profound methods of mathematical investigation,' this work is well adapted."—*Chemical News.*

HANDBOOK OF HEAT. Edited and almost entirely Re-written by BENJAMIN LOEWY, F.R.A.S., &c. Post 8vo, 6s. cloth.
"The style is always clear and precise, and conveys instruction without leaving any cloudiness or lurking doubts behind."—*Engineering.*

HANDBOOK OF OPTICS. By Dr. LARDNER. Edited by T. O. HARDING, B.A. Post 8vo, 5s. cloth.
"Written by an able scientific writer and beautifully illustrated."—*Mechanic's Magazine.*

HANDBOOK OF ELECTRICITY AND MAGNETISM. By Dr. LARDNER. Edited by G. C. FOSTER, B.A. Post 8vo, 5s. cloth.
"The book could not have been entrusted to anyone better calculated to preserve the terse and lucid style of Lardner."—*Popular Science Review.*

HANDBOOK OF ASTRONOMY. By Dr. LARDNER. Fourth Edition by E. DUNKIN, F.R.A.S. Post 8vo, 9s. 6d. cloth.
"Probably no other book contains the same amount of information in so compendious and well-arranged a form—certainly none at the price at which this is offered to the public."—*Athenæum.*
"We can do no other than pronounce this work a most valuable manual of astronomy, and we strongly recommend it to all who wish to acquire a general—but at the same time correct—acquaintance with this sublime science."—*Quarterly Journal of Science.*

DR. LARDNER'S MUSEUM OF SCIENCE AND ART.

THE MUSEUM OF SCIENCE AND ART. Edited by Dr. LARDNER. With upwards of 1,200 Engravings on Wood. In 6 Double Volumes, £1 1s. in a new and elegant cloth binding; or handsomely bound in half-morocco, 31s. 6d.
"A cheap and interesting publication, alike informing and attractive. The papers combine subjects of importance and great scientific knowledge, considerable inductive powers, and a popular style of treatment."—*Spectator.*
The 'Museum of Science and Art' is the most valuable contribution that has ever been made to the Scientific Instruction of every class of society."—Sir DAVID BREWSTER, in the *North British Review.*

*** *Separate books formed from the above, fully Illustrated, suitable for Workmen's Libraries, Science Classes, etc.*

Common Things Explained. 5s.	**Steam and its Uses.** 2s. cloth.
The Microscope. 2s. cloth.	**Popular Astronomy.** 4s. 6d. cloth.
Popular Geology. 2s. 6d. cloth.	**The Bee and White Ants.** 2s. cloth.
Popular Physics. 2s. 6d. cloth.	**The Electric Telegraph.** 1s.

Dr. Lardner's School Handbooks.

NATURAL PHILOSOPHY FOR SCHOOLS. Fcap. 8vo, 3s. 6d.
"A very convenient class-book for junior students in private schools."—*British Quarterly Review.*

ANIMAL PHYSIOLOGY FOR SCHOOLS. Fcap. 8vo, 3s. 6d.
"Clearly written, well arranged, and excellently illustrated."—*Gardener's Chronicle.*

THE ELECTRIC TELEGRAPH. By Dr. LARDNER. Revised by E. B. BRIGHT, F.R.A.S. Fcap. 8vo, 2s. 6d. cloth.
"One of the most readable books extant on the Electric Telegraph."—*English Mechanic.*

CHEMICAL MANUFACTURES, CHEMISTRY.

Chemistry for Engineers, etc.

ENGINEERING CHEMISTRY: A Practical Treatise for the Use of Analytical Chemists, Engineers, Iron Masters, Iron Founders, Students, and others. Comprising Methods of Analysis and Valuation of the Principal Materials used in Engineering Work, with numerous Analyses, Examples, and Suggestions. By H. JOSHUA PHILLIPS, F.I.C., F.C.S. formerly Analytical and Consulting Chemist to the Great Eastern Railway. Second Edition, Revised and Enlarged. Crown 8vo, 400 pp., with Illustrations, 10s. 6d. cloth. [*Just published.*

"In this work the author has rendered no small service to a numerous body of practical men. . . . The analytical methods may be pronounced most satisfactory, being as accurate as the despatch required of engineering chemists permits."—*Chemical News.*

"The book will be very useful to those who require a handy and concise *resumé* of approved methods of analysing and valuing metals, oils, fuels, &c. It is, in fact, a work for chemists, a guide to the routine of the engineering laboratory. . . . The book is full of good things. As a handbook of technical analysis, it is very welcome."—*Builder.*

"The analytical methods given are, as a whole, such as are likely to give rapid and trustworthy results in experienced hands. There is much excellent descriptive matter in the work, the chapter on 'Oils and Lubrication' being specially noticeable in this respect."—*Engineer.*

Explosives and Dangerous Goods.

DANGEROUS GOODS: Their Sources and Properties, Modes of Storage, and Transport. With Notes and Comments on Accidents arising therefrom, together with the Government and Railway Classifications, Acts of Parliament, &c. A Guide for the use of Government and Railway Officials, Steamship Owners, Insurance Companies and Manufacturers and users of Explosives and Dangerous Goods. By H. JOSHUA PHILLIPS, F.I.C., F.C.S., Author of "Engineering Chemistry, &c." Crown 8vo, 350 pp., 9s. cloth. [*Just ready.*

Alkali Trade, Manufacture of Sulphuric Acid, etc.

A MANUAL OF THE ALKALI TRADE, including the Manufacture of Sulphuric Acid, Sulphate of Soda, and Bleaching Powder. By JOHN LOMAS, Alkali Manufacturer, Newcastle-upon-Tyne and London. With 232 Illustrations and Working Drawings, and containing 390 pages of Text. Second Edition, with Additions. Super-royal 8vo, £1 10s. cloth.

"This book is written by a manufacturer for manufacturers. The working details of the most approved forms of apparatus are given, and these are accompanied by no less than 232 wood engravings, all of which may be used for the purposes of construction. Every step in the manufacture is very fully described in this manual, and each improvement explained."—*Athenæum.*

The Blowpipe.

THE BLOWPIPE IN CHEMISTRY, MINERALOGY, AND GEOLOGY. Containing all known Methods of Anhydrous Analysis, many Working Examples, and Instructions for Making Apparatus. By Lieut.-Colonel W. A. Ross, R.A., F.G.S. With 120 Illustrations. Second Edition, Revised and Enlarged. Crown 8vo, 5s. cloth.

"The student who goes through the course of experimentation here laid down will gain a better insight into inorganic chemistry and mineralogy than if he had 'got up' any of the best text-books, and passed any number of examinations in their contents."—*Chemical News.*

Commercial Chemical Analysis.

THE COMMERCIAL HANDBOOK OF CHEMICAL ANALYSIS; or, Practical Instructions for the determination of the Intrinsic or Commercial Value of Substances used in Manufactures, in Trades, and in the Arts. By A. NORMANDY. New Edition, by H. M. NOAD, F.R.S. Crown 8vo, 12s. 6d. cloth.

"We strongly recommend this book to our readers as a guide, alike indispensable to the housewife as to the pharmaceutical practitioner."—*Medical Times.*

Dye-Wares and Colours.

THE MANUAL OF COLOURS AND DYE-WARES: Their Properties, Applications, Valuations, Impurities, and Sophistications. For the use of Dyers, Printers, Drysalters, Brokers, &c. By J. W. SLATER. Second Edition, Revised and greatly Enlarged. Crown 8vo, 7s. 6d. cloth.

"A complete encyclopædia of the *materia tinctoria*. The information given respecting each article is full and precise, and the methods of determining their value are given with clearness, and are practical as well as valuable."—*Chemist and Druggist.*

CHEMICAL MANUFACTURES, CHEMISTRY, etc.   35

Modern Brewing and Malting.
A HANDYBOOK FOR BREWERS: Being a Practical Guide to the Art of Brewing and Malting. Embracing the Conclusions of Modern Research which bear upon the Practice of Brewing. By HERBERT EDWARDS WRIGHT, M.A. Crown 8vo, 550 pp., 12s. 6d, cloth.

"May be consulted with advantage by the student who is preparing himself for examinational tests, while the scientific brewer will find in it a *resume* of all the most important discoveries of modern times. The work is written throughout in a clear and concise manner, and the author takes great care to discriminate between vague theories and well-established facts."—*Brewers' Journal.*

"We have great pleasure in recommending this handybook, and have no hesitation in saying that it is one of the best—if not the best—which has yet been written on the subject of beer-brewing in this country, and it should have a place on the shelves of every brewer's library."—*The Brewer's Guardian.*

"Although the requirements of the student are primarily considered, an acquaintance of half-an-hour's duration cannot fail to impress the practical brewer with the sense of having found a trustworthy guide and practical counsellor in brewery matters."—*Chemical Trade Journal.*

Analysis and Valuation of Fuels.
FUELS: SOLID, LIQUID, AND GASEOUS, Their Analysis and Valuation. For the Use of Chemists and Engineers. By H. J. PHILLIPS, F.C.S., formerly Analytical and Consulting Chemist to the Great Eastern Railway. Second Edition, Revised and Enlarged. Crown 8vo, 5s. cloth.

"Ought to have its place in the laboratory of every metallurgical establishment, and wherever fuel is used on a large scale."—*Chemical News.*

"Cannot fail to be of wide interest, especially at the present time."—*Railway News.*

Pigments.
THE ARTIST'S MANUAL OF PIGMENTS. Showing their Composition, Conditions of Permanency, Non-Permanency, and Adulterations; Effects in Combination with Each Other and with Vehicles, and the most Reliable Tests of Purity Together with the Science and Art Department's Examination Questions on Painting. By H. C. STANDACE. Second Edition. Crown 8vo, 2s. 6d. cloth.

"This work is indeed *multum-in-parvo*, and we can, with good conscience, recommend it to all who come in contact with pigments, whether as makers, dealers or users."—*Chemical Review.*

Gauging. Tables and Rules for Revenue Officers, Brewers, etc.
A POCKET BOOK OF MENSURATION AND GAUGING: Containing Tables, Rules and Memoranda for Revenue Officers, Brewers, Spirit Merchants, &c. By J. B. MANT (Inland Revenue). Second Edition, Revised. 18mo, 4s. leather.

"This handy and useful book is adapted to the requirements of the Inland Revenue Department, and will be a favourite book of reference."—*Civilian.*

"Should be in the hands of every practical brewer."—*Brewers' Journal.*

INDUSTRIAL ARTS, TRADES, AND MANUFACTURES.

Cotton Spinning.
COTTON MANUFACTURE: A Manual of Practical Instruction in the Processes of Opening, Carding, Combing, Drawing, Doubling and Spinning of Cotton, the Methods of Dyeing, &c. For the Use of Operatives, Overlookers and Manufacturers. By JOHN LISTER, Technical Instructor, Pendleton. 8vo, 7s. 6d. cloth. [*Just published.*

"This invaluable volume is a distinct advance in the literature of cotton manufacture."—*Machinery.*

"It is thoroughly reliable, fulfilling nearly all the requirements desired."—*Glasgow Herald.*

Flour Manufacture, Milling, etc.
FLOUR MANUFACTURE: A Treatise on Milling Science and Practice. By FRIEDRICH KICK, Imperial Regierungsrath, Professor of Mechanical Technology in the Imperial German Polytechnic Institute, Prague. Translated from the Second Enlarged and Revised Edition with Supplement. By H. H. P. POWLES, A.-M.Inst.C.E. Nearly 400 pp. Illustrated with 28 Folding Plates, and 167 Woodcuts. Royal 8vo, 25s. cloth.

"This valuable work is, and will remain, the standard authority on the science of milling. . . . The miller who has read and digested this work will have laid the foundation, so to speak, of a successful career; he will have acquired a number of general principles which he can proceed to apply. In this handsome volume we at last have the accepted text-book of modern milling in good, sound English, which has little, if any, trace of the German idiom."—*The Miller.*

"The appearance of this celebrated work in English is very opportune, and British millers will, we are sure, not be slow in availing themselves of its pages."—*Millers' Gazette.*

Agglutinants.

CEMENTS, PASTES, GLUES AND GUMS: A Practical Guide to the Manufacture and Application of the various Agglutinants required in the Building, Metal-Working, Wood-Working and Leather-Working Trades, and for Workshop, Laboratory or Office Use. With upwards of 900 Recipes and Formulæ. By H. C. STANDAGE, Chemist. Crown 8vo, 2s. 6d. cloth. [*Just published.*

"We have pleasure in speaking favourably of this volume. So far as we have had experience, which is not inconsiderable, this manual is trustworthy."—*Athenæum.*

"As a revelation of what are considered trade secrets, this book will arouse an amount of curiosity among the large number of industries it touches."—*Daily Chronicle.*

"In this goodly collection of receipts it would be strange if a cement for any purpose cannot be found."—*Oil and Colourman's Journal.*

Soap-making.

THE ART OF SOAP-MAKING: A Practical Handbook of the Manufacture of Hard and Soft Soaps, Toilet Soaps, etc. Including many New Processes, and a Chapter on the Recovery of Glycerine from Waste Leys. By ALEXANDER WATT. Fourth Edition, Enlarged. Crown 8vo, 7s. 6d. cloth.

"The work will prove very useful, not merely to the technological student, but to the practical soap-boiler who wishes to understand the theory of his art."—*Chemical News.*

"A thoroughly practical treatise on an art which has almost no literature in our language. We congratulate the author on the success of his endeavour to fill a void in English technical literature."—*Nature.*

Paper Making.

PRACTICAL PAPER-MAKING: A Manual for Paper-makers and Owners and Managers of Paper-Mills. With Tables, Calculations, &c. By G. CLAPPERTON, Paper-maker. With Illustrations of Fibres from Micro-Photographs. Crown 8vo, 5s. cloth. [*Just published.*

"The author caters for the requirements of responsible mill hands, apprentices, &c., whilst his manual will be found of great service to students of technology, as well as to veteran paper makers and mill owners. The illustrations form an excellent feature."—*Paper Trade Review.*

"We recommend everybody interested in the trade to get a copy of this thoroughly practical book."—*Paper Making.*

Paper Making.

THE ART OF PAPER MAKING: A Practical Handbook of the Manufacture of Paper from Rags, Esparto, Straw, and other Fibrous Materials, Including the Manufacture of Pulp from Wood Fibre, with a Description of the Machinery and Appliances used. To which are added Details of Processes for Recovering Soda from Waste Liquors. By ALEXANDER WATT, Author of "The Art of Soap-Making." With Illusts. Crown 8vo, 7s. 6d. cloth.

"It may be regarded as the standard work on the subject. The book is full of valuable information. The 'Art of Paper-making,' is in every respect a model of a text-book, either for a technical class or for the private student."—*Paper and Printing Trades Journal.*

Leather Manufacture.

THE ART OF LEATHER MANUFACTURE. Being a Practical Handbook, in which the Operations of Tanning, Currying, and Leather Dressing are fully Described, and the Principles of Tanning Explained, and many Recent Processes Introduced; as also the Methods for the Estimation of Tannin, and a Description of the Arts of Glue Boiling, Gut Dressing, &c. By ALEXANDER WATT, Author of "Soap-Making," &c. Second Edition. Crown 8vo, 9s. cloth.

"A sound, comprehensive treatise on tanning and its accessories. It is an eminently valuable production, which redounds to the credit of both author and publishers."—*Chemical Review.*

Boot and Shoe Making.

THE ART OF BOOT AND SHOE-MAKING. A Practical Handbook, including Measurement, Last-Fitting, Cutting-Out, Closing, and Making, with a Description of the most approved Machinery employed. By JOHN B. LENO, late Editor of *St. Crispin*, and *The Boot and Shoe-Maker*. 12mo, 2s. cloth limp.

"This excellent treatise is by far the best work ever written. The chapter on clicking, which shows how waste may be prevented, will save fifty times the price of the book."
—*Scottish Leather Trader.*

Wood Engraving.

WOOD ENGRAVING: *A Practical and Easy Introduction to the Study of the Art.* By WILLIAM NORMAN BROWN. Second Edition. With numerous Illustrations. 12mo, 1s. 6d. cloth limp.

"The book is clear and complete, and will be useful to anyone wanting to understand the first elements of the beautiful art of wood engraving."—*Graphic.*

Watch Adjusting.

THE WATCH ADJUSTER'S MANUAL: A Practical Guide for the Watch and Chronometer Adjuster in Making, Springing, Timing and Adjusting for Isochronism, Positions and Temperatures. By C. E. FRITTS. 370 pages, with Illustrations, 8vo, 16s. cloth. [*Just published.*

Horology.

A TREATISE ON MODERN HOROLOGY, in *Theory and Practice.* Translated from the French of CLAUDIUS SAUNIER, ex-Director of the School of Horology at Maçon, by JULIEN TRIPPLIN, F.R.A.S., Besançon Watch Manufacturer, and EDWARD RIGG, M.A., Assayer in the Royal Mint. With 78 Woodcuts and 22 Coloured Copper Plates. Second Edition. Super-royal 8vo, £2 2s. cloth; £2 10s. half-calf.

"There is no horological work in the English language at all to be compared to this production of M. Saunier's for clearness and completeness. It is alike good as a guide for the student and as a reference for the experienced horologist and skilled workman."—*Horological Journal.*

"The latest, the most complete, and the most reliable of those literary productions to which continental watchmakers are indebted for the mechanical superiority over their English brethren—in fact, the Book of Books, is M. Saunier's ' Treatise.'"—*Watchmaker, Jeweller and Silversmith.*

Watchmaking.

THE WATCHMAKER'S HANDBOOK. Intended as a Workshop Companion for those engaged in Watchmaking and the Allied Mechanical Arts. Translated from the French of CLAUDIUS SAUNIER, and considerably enlarged by JULIEN TRIPPLIN, F.R.A.S., Vice-President of the Horological Institute, and EDWARD RIGG, M.A., Assayer in the Royal Mint. With numerous Woodcuts and 14 Copper Plates. Third Edition. Crown 8vo, 9s. cloth.

"Each part is truly a treatise in itself. The arrangement is good and the language is clear and concise. It is an admirable guide for the young watchmaker."—*Engineering.*

"It is impossible to speak too highly of its excellence. It fulfils every requirement in a handbook intended for the use of a workman. Should be found in every workshop."—*Watch and Clockmaker.*

"This book contains an immense number of practical details bearing on the daily occupation of a watchmaker."—*Watchmaker and Metalworker* (Chicago).

Watches and Timekeepers.

A HISTORY OF WATCHES AND OTHER TIMEKEEPERS. By JAMES F. KENDAL, M.B.H.Inst. 1s. 6d. boards; or 2s. 6d. cloth gilt.

"Mr. Kendal's book, for its size, is the best which has yet appeared on this subject in the English language."—*Industries.*

"Open the book where you may, there is interesting matter in it concerning the ingenious devices of the ancient or modern horologer. The subject is treated in a liberal and entertaining spirit, as might be expected of a historian who is a master of the craft."—*Saturday Review.*

Electrolysis of Gold, Silver, Copper, etc.

ELECTRO-DEPOSITION: *A Practical Treatise on the Electrolysis of Gold, Silver, Copper, Nickel, and other Metals and Al oys.* With descriptions of Voltaic Batteries, Magneto and Dynamo-Electric Machines, Thermopiles, and of the Materials and Processes used in every Department of the Art, and several Chapters on Electro-Metallurgy. By ALEXANDER WATT, Author of "Electro-Metallurgy," &c. Third Edition, Revised. Crown 8vo, 9s. cloth.

"Eminently a book for the practical worker in electro-deposition. It contains practical descriptions of methods, processes and materials as actually pursued and used in the workshop."—*Engineer.*

Electro-Metallurgy.

ELECTRO-METALLURGY; *Practically Treated.* By ALEXANDER WATT, Author of "Electro-Deposition," &c. Tenth Edition, including the most recent Processes. 12mo, 4s. cloth boards.

"From this book both amateur and artisan may learn everything necessary for the successful prosecution of electroplating."—*Iron.*

Working in Gold.

THE JEWELLER'S ASSISTANT IN THE ART OF WORKING IN GOLD: A Practical Treatise for Masters and Workmen. Compiled from the Experience of Thirty Years' Workshop Practice. By GEORGE E. GEE, Author of "The Goldsmith's Handbook," &c. Cr. 8vo, 7s. 6d. cloth.

"This manual of technical education is apparently destined to be a valuable auxiliary to a handicraft which is certainly capable of great improvement."—*The Times.*

"Very useful in the workshop, as the knowledge is practical, having been acquired by long experience, and all the recipes and directions are guaranteed to be successful."—*Jeweller and Metalworker.*

Electroplating.

ELECTROPLATING: A Practical Handbook on the Deposition of Copper, Silver, Nickel, Gold, Aluminium, Brass, Platinum, &c. &c. With Descriptions of the Chemicals, Materials, Batteries, and Dynamo Machines used in the Art. By J. W. URQUHART, C.E., Author of "Electric Light," &c. Third Edition, with Additions. Crown 8vo. 5s. cloth.

"An excellent practical manual."—*Engineering.*
"An excellent work, giving the newest information."—*Horological Journal.*

Electrotyping.

ELECTROTYPING: *The Reproduction and Multiplication of Printing Surfaces and Works of Art by the Electro-deposition of Metals.* By J. W. URQUHART, C.E. Crown 8vo, 5s. cloth.

"The book is thoroughly practical. The reader is, therefore, conducted through the leading laws of electricity, then through the metals used by electrotypers, the apparatus, and the depositing processes, up to the final preparation of the work."—*Art Journal.*

Goldsmiths' Work.

THE GOLDSMITH'S HANDBOOK. By GEORGE E. GEE, Jeweller, &c. Third Edition, considerably Enlarged. 12mo, 3s. 6d. cl. bds.

"A good, sound educator, which will be accepted as an authority."—*Horological Journal.*

Silversmiths' Work.

THE SILVERSMITH'S HANDBOOK. By GEORGE E. GEE, Jeweller, &c. Second Edition, Revised. 12mo, 3s. 6d. cloth.

"The chief merit of the work is its practical character. . . The workers in the trade will speedily discover its merits when they sit down to study it."—*English Mechanic.*

** The above two works together, strongly half-bound, price 7s.

Sheet Metal Working.

THE SHEET METAL WORKER'S INSTRUCTOR: For Zinc, Sheet Iron, Copper, and Tin Plate Workers. Containing Rules for describing the Patterns required in the Different Branches of the Trade. By R. H. WARN, Tin Plate Worker. With Thirty-two Plates. 8vo, 7s. 6d. cl.

Bread and Biscuit Baking.

THE BREAD AND BISCUIT BAKER'S AND SUGAR-BOILER'S ASSISTANT. Including a large variety of Modern Recipes. By ROBERT WELLS, Practical Baker. Crown 8vo, 2s. cloth.

"A large number of wrinkles for the ordinary cook, as well as the baker."—*Saturday Review.*

Confectionery for Hotels and Restaurants.

THE PASTRYCOOK AND CONFECTIONER'S GUIDE. For Hotels, Restaurants and the Trade in general, adapted also for Family Use. By ROBERT WELLS. Crown 8vo, 2s. cloth.

"We cannot speak too highly of this really excellent work. In these days of keen competition our readers cannot do better than purchase this book."—*Bakers' Times.*

Ornamental Confectionery.

ORNAMENTAL CONFECTIONERY: A Guide for Bakers, Confectioners and Pastrycooks; including a variety of Modern Recipes, and Remarks on Decorative and Coloured Work. With 129 Original Designs. By ROBERT WELLS. Crown 8vo, cloth gilt, 5s.

"A valuable work, practical, and should be in the hands of every baker and confectioner. The illustrative designs are alone worth treble the amount charged for the whole work."—*Bakers' Times.*

Flour Confectionery.

THE MODERN FLOUR CONFECTIONER. Wholesale and Retail. Containing a large Collection of Recipes for Cheap Cakes, Biscuits, &c. With Remarks on the Ingredients used in their Manufacture. By R. WELLS. Crown 8vo, 2s. cloth

"The work is of a decidedly practical character, and in every recipe regard is had to economical working."—*North British Daily Mail.*

Laundry Work.

LAUNDRY MANAGEMENT. A Handbook for Use in Private and Public Laundries, Including Descriptive Accounts of Modern Machinery and Appliances for Laundry Work. By the EDITOR of "The Laundry Journal." Second Edition. Crown 8vo, 2s. 6d. cloth.

"This book should certainly occupy an honoured place on the shelves of all housekeepers who wish to keep themselves *au courant* of the newest appliances and methods."—*The Queen.*

HANDYBOOKS FOR HANDICRAFTS.
By PAUL N. HASLUCK,
Editor of "Work" (New Series); Author of "Lathework," "Milling Machines," &c.

Crown 8vo, 144 pages, cloth, price 1s. each.

☞ *These* Handybooks *have been written to supply information for* Workmen, Students, *and* Amateurs *in the several Handicrafts, on the actual* Practice *of the* Workshop, *and are intended to convey in plain language* Technical Knowledge *of the several* Crafts. *In describing the processes employed, and the manipulation of material, workshop terms are used; workshop practice is fully explained; and the text is freely illustrated with drawings of modern tools, appliances, and processes.*

THE METAL TURNER'S HANDYBOOK. A Practical Manual for Workers at the Foot-Lathe. With over 100 Illustrations. Price 1s.
"The book will be of service alike to the amateur and the artisan turner. It displays thorough knowledge of the subject."—*Scotsman.*

THE WOOD TURNER'S HANDYBOOK. A Practical Manual for Workers at the Lathe. With over 100 Illustrations. Price 1s.
"We recommend the book to young turners and amateurs. A multitude of workmen have hitherto sought in vain for a manual of this special industry."—*Mechanical World.*

THE WATCH JOBBER'S HANDYBOOK. A Practical Manual on Cleaning, Repairing, and Adjusting. With upwards of 100 Illustrations. Price 1s.
"We strongly advise all young persons connected with the watch trade to acquire and study this inexpensive work."—*Clerkenwell Chronicle.*

THE PATTERN MAKER'S HANDYBOOK. A Practical Manual on the Construction of Patterns for Founders. With upwards of 100 Illustrations. Price 1s.
"A most valuable, if not indispensable, manual for the pattern maker."—*Knowledge.*

THE MECHANIC'S WORKSHOP HANDYBOOK. A Practical Manual on Mechanical Manipulation. Embracing Information on various Handicraft Processes, with Useful Notes and Miscellaneous Memoranda. Comprising about 200 Subjects. Price 1s.
"A very clever and useful book, which should be found in every workshop; and it should certainly find a place in all technical schools."—*Saturday Review.*

THE MODEL ENGINEER'S HANDYBOOK. A Practical Manual on the Construction of Model Steam Engines. With upwards of 100 Illustrations. Price 1s.
"Mr. Hasluck has produced a very good little book."—*Builder.*

THE CLOCK JOBBER'S HANDYBOOK. A Practical Manual on Cleaning, Repairing, and Adjusting. With upwards of 100 Illustrations. Price 1s.
"It is of inestimable service to those commencing the trade."—*Coventry Standard.*

THE CABINET WORKER'S HANDYBOOK: A Practical Manual on the Tools, Materials, Appliances, and Processes employed in Cabinet Work. With upwards of 100 Illustrations. Price 1s.
"Mr. Hasluck's thoroughgoing little Handybook is amongst the most practical guides we have seen for beginners in cabinet-work."—*Saturday Review.*

THE WOODWORKER'S HANDYBOOK OF MANUAL INSTRUCTION. Embracing Information on the Tools, Materials, Appliances and Processes employed in Woodworking. With 104 Illustrations. Price 1s
[*Just published.*

THE METALWORKER'S HANDYBOOK. With upwards of 100 Illustrations. [*In preparation.*

⁎ Opinions of the Press.

"Written by a man who knows, not only how work ought to be done, but how to do it, and how to convey his knowledge to others."—*Engineering.*
"Mr. Hasluck writes admirably, and gives complete instructions."—*Engineer.*
"Mr. Hasluck combines the experience of a practical teacher with the manipulative skill and scientific knowledge of processes of the trained mechanician, and the manuals are marvels of what can be produced at a popular price."—*Schoolmaster.*
"Helpful to workmen of all ages and degrees of experience."—*Daily Chronicle.*
"Practical, sensible, and remarkably cheap."—*Journal of Education.*
"Concise, clear and practical."—*Saturday Review.*

COMMERCE, COUNTING-HOUSE WORK, TABLES, etc.

Commercial French.

A NEW BOOK OF COMMERCIAL FRENCH: Grammar—Vocabulary — Correspondence — Commercial Documents — Geography — Arithmetic—Lexicon. By P. CARROUÉ, Professor in the City High School J.—B. Say (Paris). Crown 8vo, 4s. 6d. cloth. [*Just published*

Commercial Education.

LESSONS IN COMMERCE. By Professor R. GAMBARO, of the Royal High Commercial School at Genoa. Edited and Revised by JAMES GAULT, Professor of Commerce and Commercial Law in King's College, London. Second Edition, Revised. Crown 8vo, 3s. 6d. cloth. [*Just published.*

"The publishers of this work have rendered considerable service to the cause of commercial education by the opportune production of this volume. . . . The work is peculiarly acceptable to English readers and an admirable addition to existing class-books. In a phrase, we think the work attains its object in furnishing a brief account of those laws and customs of British trade with which the commercial man interested therein should be familiar."—*Chamber of Commerce Journal*

"An invaluable guide in the hands of those who are preparing for a commercial career."—*Counting House*

Foreign Commercial Correspondence.

THE FOREIGN COMMERCIAL CORRESPONDENT: Being Aids to Commercial Correspondence in Five Languages—English, French, German, Italian, and Spanish. By CONRAD E. BAKER. Second Edition. Crown 8vo, 3s. 6d. cloth.

"Whoever wishes to correspond in all the languages mentioned by Mr. Baker cannot do better than study this work, the materials of which are excellent and conveniently arranged. They consist not of entire specimen letters but—what are far more useful—short passages, sentences, or phrases expressing the same general idea in various forms."—*Athenæum.*

"A careful examination has convinced us that it is unusually complete, well arranged, and reliable. The book is a thoroughly good one."—*Schoolmaster.*

Accounts for Manufacturers.

FACTORY ACCOUNTS: Their Principles and Practice. A Handbook for Accountants and Manufacturers, with Appendices on the Nomenclature of Machine Details; the Income Tax Acts; the Rating of Factories; Fire and Boiler Insurance; the Factory and Workshop Acts, &c., including also a Glossary of Terms and a large number of Specimen Rulings. By EMILE GARCKE and J. M. FELLS. Fourth Edition, Revised and Enlarged. Demy 8vo, 250 pages, 6s. strongly bound.

"A very interesting description of the requirements of Factory Accounts. . . . the principle of assimilating the Factory Accounts to the general commercial books is one which we thoroughly agree with."—*Accountants' Journal.*

"Characterised by extreme thoroughness. There are few owners of factories who would not derive great benefit from the perusal of this most admirable work."—*Local Government Chronicle.*

Modern Metrical Units and Systems.

MODERN METROLOGY: A Manual of the Metrical Units and Systems of the Present Century. With an Appendix containing a proposed English System. By LOWIS D'A. JACKSON, A.M.Inst.C.E., Author of "Aid to Survey Practice," &c. Large crown 8vo, 12s. 6d. cloth.

"We recommend the work to all interested in the practical reform of our weights and measures."—*Nature.*

The Metric System and the British Standards.

A SERIES OF METRIC TABLES, in which the British Standard Measures and Weights are compared with those of the Metric System at present in Use on the Continent. By C. H. DOWLING, C.E. 8vo, 10s. 6d. strongly bound

"Mr. Dowling's Tables are well put together as a ready-reckoner for the conversion of one system into the other."—*Athenæum.*

Iron Shipbuilders' and Merchants' Weight Tables.

IRON-PLATE WEIGHT TABLES: For Iron Shipbuilders, Engineers, and Iron Merchants. Containing the Calculated Weights of upwards of 150,000 different sizes of Iron Plates, from 1 foot by 6 in. by ¼ in. to 10 feet by 5 feet by 1 in. Worked out on the basis of 40 lbs. to the square foot of Iron of 1 inch in thickness. Carefully compiled and thoroughly Revised by H. BURLINSON and W. H. SIMPSON. Oblong 4to, 25s. half-bound.

"This work will be found of great utility. The authors have had much practical experience of what is wanting in making estimates; and the use of the book will save much time in making elaborate calculations."—*English Mechanic.*

Chadwick's Calculator for Numbers and Weights Combined.

THE NUMBER, WEIGHT, AND FRACTIONAL CALCULATOR. Containing upwards of 250,000 Separate Calculations, showing at a glance the value at 422 different rates, ranging from $\frac{1}{16}$th of a Penny to 20s. each, or per cwt., and £20 per ton, of any number of articles consecutively, from 1 to 470.—Any number of cwts., qrs., and lbs., from 1 cwt. to 470 cwts.—Any number of tons, cwts., qrs., and lbs., from 1 to 1,000 tons. By WILLIAM CHADWICK, Public Accountant. Third Edition, Revised and Improved. 8vo, 18s., strongly bound for Office wear and tear.

☞ Is adapted for the use of Accountants and Auditors, Railway Companies, Canal Companies, Shippers, Shipping Agents, General Carriers, etc. Ironfounders, Brassfounders, Metal Merchants, Iron Manufacturers, Ironmongers, Engineers, Machinists, Boiler Makers, Millwrights, Roofing, Bridge and Girder Makers, Colliery Proprietors, etc. Timber Merchants, Builders, Contractors, Architects, Surveyors, Auctioneers, Valuers, Brokers, Mill Owners and Manufacturers, Mill Furnishers, Merchants, and General Wholesale Tradesmen. Also for the Apportionment o Mileage Charges for Railway Traffic.

"It is as easy of reference for any answer or any number of answers as a dictionary, and the references are even more quickly made. For making up accounts or estimates the book must prove invaluable to all who have any considerable quantity of calculations involving price and measure in any combination to do."—*Engineer.*

Harben's Comprehensive Weight Calculator.

THE WEIGHT CALCULATOR. Being a Series of Tables upon a New and Comprehensive Plan, exhibiting at One Reference the exact Value of any Weight from 1 lb. to 15 tons, at 300 Progressive Rates, from 1d. to 168s. per cwt., and containing 186,000 Direct Answers, which, with their Combinations, consisting of a single addition (mostly to be performed at sight), will afford an aggregate of 10,266,000 Answers; the whole being calculated and designed to ensure correctness and promote despatch. By HENRY HARBEN, Accountant. Fourth Edition, carefully Corrected. Royal 8vo, £1 5s. strongly half-bound.

"A practical and useful work of reference for men of business generally; it is the best of the kind we have seen."—*Ironmonger.*

"Of priceless value to business men. It is a necessary book in all mercantile offices."—*Sheffield Independent.*

Harben's Comprehensive Discount Guide.

THE DISCOUNT GUIDE. Comprising several Series of Tables for the use of Merchants, Manufacturers, Ironmongers, and others, by which may be ascertained the exact Profit arising from any mode of using Discounts, either in the Purchase or Sale of Goods, and the method of either Altering a Rate of Discount or Advancing a Price, so as to produce, by one operation, a sum that will realise any required profit after allowing one or more Discounts: to which are added Tables of Profit or Advance from 1¼ to 90 per cent., Tables of Discount from 1¼ to 98¾ per cent., and Tables of Commission, &c., from ⅛ to 10 per cent. By HENRY HARBEN, Accountant. New Edition, Revised and Corrected. Demy 8vo, 544 pp., £1 5s. half-bound.

"A book such as this can only be appreciated by business men, to whom the saving of time means saving of money. We have the high authority of Professor J. R. Young that the tables throughout the work are constructed upon strictly accurate principles. The work is a model of typographical clearness, and must prove of great value to merchants, manufacturers, and general traders."—*British Trade Journal.*

New Wages Calculator.

TABLES OF WAGES at 54, 52, 50 and 48 Hours per Week. Showing the Amounts of Wages from One-quarter-of-an hour to Sixty-four hours in each case at Rates of Wages advancing by One Shilling from 4s. to 55s. per week. By THOS. GARBUTT, Accountant. Square crown 8vo, 6s. half-bound. [*Just published.*

Iron and Metal Trades' Calculator.

THE IRON AND METAL TRADES' COMPANION. For expeditiously ascertaining the Value of any Goods bought or sold by Weight, from 1s. per cwt. to 112s. per cwt., and from one farthing per pound to one shilling per pound. By THOMAS DOWNIE. 396 pp., 9s. leather.

"A most useful set of tables; nothing like them before existed."—*Building News.*

"Although specially adapted to the iron and metal trades, the tables will be found useful in every other business in which merchandise is bought and sold by weight."—*Railway News.*

"DIRECT CALCULATORS,"
By M. B. COTSWORTH, of Holgate, York.

**QUICKEST AND MOST ACCURATE MEANS OF CALCULATION KNOWN.
ENSURE ACCURACY and SPEED WITH EASE, SAVE TIME and MONEY.**
Accounts may be charged out or checked by these means in about one third he time required by ordinary methods of calculation. These unrivalled "Calculators" have very clear and original contrivances for instantly finding the exact answer, by its fixed position, without even sighting the top or side of the page. They are varied in arrangement to suit the special need of each particular trade.

All the leading firms now use Calculators, even where they employ experts.

N.B.—Indicator letters in brackets should be quoted.

"*RAILWAY & TRADERS' CALCULATOR*" **(R. & T.)** 10s. 6d. Including Scale of Charges for Small Parcels by Merchandise Trains. "Direct Calculator"—the only Calculator published giving exact charge for Cwts., Qrs. and Lbs., together. "Calculating Tables" for every 1d. rate to 100s. per ton. "Wages Calculator." "Percentage Rates." "Grain, Flour, Ale, &c., Weight Calculators."

"*DIRECT CALCULATOR* **(I R)**" including all the above except "Calculating Tables." 7s.

"*DIRECT CALCULATOR* **(A)**" by ½d., 2s. each opening, exact pence to 40s. per ton. 5s.

"*DIRECT CALCULATOR* **(B)**" by 1d., 4s. each opening, exact pence to 40s. per ton. 4s. 6d.

"*DIRECT CALCULATOR* **(C)**" by 1d. (with Cwts. and Qrs. to nearest farthing), to 40s. per ton. 4s. 6d.

"*DIRECT CALCULATOR* **(Ds)**" by 1d. gradations. (Single Tons to 50 Tons, then by fifties to 1,000 Tons, with Cwts. values below in exact pence payable, fractions of ½d. and upwards being counted as 1d. 6s. 6d.

"*DIRECT CALCULATOR* **(D)**" has from 1,000 to 10,000 Tons in addition to the (**Ds**) Calculator. 7s. 6d.

"*DIRECT CALCULATOR* **(Es)**" by 1d. gradations. (As (**D**) to 1,000 Tons, with Cwts. and Qrs. values shown separately to the nearest farthing). 5s. 6d.

"*DIRECT CALCULATOR* **(E)**" has from 1,000 to 10,000 Tons in addition to the (**Es**) Calculator. 6s. 6d.

"*DIRECT CALCULATOR* **(F)**" by 1d., 2s. each opening, exact pence to 40s. per ton. 4s. 6d.

"*DIRECT CALCULATOR* **(G)**" by 1d., 1s. each opening; 6 in. by 9 in. Nearest ¼d. Indexed (**G I**) 3s. 6d. 2s. 6d.

"*DIRECT CALCULATOR* **(H)**" by 1d., 1s. each opening; 6 in. by 9 in. To exact pence. Indexed (**H I**) 3s. 6d. 2s. 6d.

"*DIRECT CALCULATOR* **(K)**" Showing Values of Tons, Cwts. and Qrs. in even pence (fractions of 1d. as 1d.), for the Retail Coal Trade. 4s. 6d.

"*RAILWAY AND TIMBER TRADES MEASURER AND CALCULATOR* **(T)**" (as prepared for the Railway Companies). The only book published giving true content of unequal sided and round timber by eighths of an inch, quarter girth, Weights from Cubic Feet—Standards, Superficial Feet, and Stone to Weights—Running Feet from lengths of Deals—Standard Multipliers—Timber Measures—Customs Regulations, &c. 3s. 6d.

AGRICULTURE, FARMING, GARDENING, etc.

Dr. Fream's New Edition of "The Standard Treatise on Agriculture."

THE COMPLETE GRAZIER, and FARMER'S and CATTLE-BREEDER'S ASSISTANT: A Compendium of Husbandry. Originally Written by WILLIAM YOUATT. Thirteenth Edition, entirely Re-written, considerably Enlarged, and brought up to the Present Requirements of Agricultural Practice, by WILLIAM FREAM, LL.D., Steven Lecturer in the University of Edinburgh, Author of "The Elements of Agriculture," &c. Royal 8vo, 1,100 pp., with over 450 Illustrations. £1 11s. 6d. strongly and handsomely bound.

EXTRACT FROM PUBLISHERS' ADVERTISEMENT.

"A treatise that made its original appearance in the first decade of the century, and that enters upon its Thirteenth Edition before the century has run its course, has undoubtedly established its position as a work of permanent value. . . The phenomenal progress of the last dozen years in the Practice and Science of Farming has rendered it necessary, however, that the volume should be re-written, . . . and for this undertaking the publishers were fortunate enough to secure the services of Dr. FREAM, whose high attainments in all matters pertaining to agriculture have been so emphatically recognised by the highest professional and official authorities. In carrying out his editorial duties, Dr. FREAM has been favoured with valuable contributions by Prof. J WORTLEY AXE, Mr. E. BROWN, Dr. BERNARD DYER, Mr. W. J. MALDEN, Mr. R. H. REW Prof. SHELDON, Mr. J. SINCLAIR, Mr. SANDERS SPENCER, and others.

"As regards the illustrations of the work, no pains have been spared to make them as representative and characteristic as possible, so as to be practically useful to the Farmer and Grazier."

SUMMARY OF CONTENTS.

BOOK I. ON THE VARIETIES, BREEDING, REARING, FATTENING, AND MANAGEMENT OF CATTLE.
BOOK II. ON THE ECONOMY AND MANAGEMENT OF THE DAIRY.
BOOK III. ON THE BREEDING, REARING, AND MANAGEMENT OF HORSES.
BOOK IV. ON THE BREEDING, REARING, AND FATTENING OF SHEEP.
BOOK V. ON THE BREEDING, REARING, AND FATTENING OF SWINE.
BOOK VI. ON THE DISEASES OF LIVE STOCK.
BOOK VII. ON THE BREEDING, REARING, AND MANAGEMENT OF POULTRY.
BOOK VIII. ON FARM OFFICES AND IMPLEMENTS OF HUSBANDRY.
BOOK IX. ON THE CULTURE AND MANAGEMENT OF GRASS LANDS.
BOOK X. ON THE CULTIVATION AND APPLICATION OF GRASSES, PULSE, AND ROOTS.
BOOK XI. ON MANURES AND THEIR APPLICATION TO GRASS LAND & CROPS
BOOK XII. MONTHLY CALENDARS OF FARMWORK.

** OPINIONS OF THE PRESS ON THE NEW EDITION.

"Dr. Fream is to be congratulated on the successful attempt he has made to give us a work which will at once become the standard classic of the farm practice of the country. We believe that it will be found that it has no compeer among the many works at present in existence. . . . The illustrations are admirable, while the frontispiece, which represents the well-known bull, New Year's Gift, bred by the Queen, is a work of art."—*The Times.*

"The book must be recognised as occupying the proud position of the most exhaustive work freference in the English language on the subject with which it deals."—*Athenæum.*

"The most comprehensive guide to modern farm practice that exists in the English language to-day. . . . The book is one that ought to be on every farm and in the library of every landowner."—*Mark Lane Express.*

"In point of exhaustiveness and accuracy the work will certainly hold a pre-eminent and unique position among books dealing with scientific agricultural practice. It is, in fact, an agricultural library of itself."—*North British Agriculturist.*

"A compendium of authoritative and well-ordered knowledge on every conceivable branch of the work of the live stock farmer; probably without an equal in this or any other country."
Yorkshire Post.

British Farm Live Stock.

FARM LIVE STOCK OF GREAT BRITAIN. By ROBERT WALLACE, F.L.S., F.R.S.E., &c., Professor of Agriculture and Rural Economy in the University of Edinburgh. Third Edition, thoroughly Revised and considerably Enlarged. With over 120 Phototypes of Prize Stock. Demy 8vo, 384 pp., with 79 Plates and Maps, 12s. 6d. cloth.

"A really complete work on the history, breeds, and management of the farm stock of Great Britain, and one which is likely to find its way to the shelves of every country gentleman's library."—*The Times.*

"The latest edition of 'Farm Live Stock of Great Britain' is a production to be proud of, and its issue not the least of the services which its author has rendered to agricultural science."
Scottish Farmer,

"The book is very attractive . . . and we can scarcely imagine the existence of a farmer who would not like to have a copy of this beautiful work."—*Mark Lane Express.*

"A work which will long be regarded as a standard authority whenever a concise history and description of the breeds of live stock in the British Isles is required."—*Bell's Weekly Messenger.*

Dairy Farming.

BRITISH DAIRYING. A Handy Volume on the Work of the Dairy-Farm. For the Use of Technical Instruction Classes, Students in Agricultural Colleges, and the Working Dairy-Farmer. By Prof. J. P. SHELDON, ate Special Commissioner of the Canadian Government, Author of "Dairy Farming," &c. With numerous Illustrations. Crown 8vo, 2s. 6d. cloth

"We confidently recommend it as a text-book on dairy farming."—*Agricultural Gazette*.
"Probably the best half-crown manual on dairy work that has yet been produced."—*North British Agriculturist.*
"It is the soundest little work we have yet seen on the subject."—*The Times.*

Dairy Manual.

MILK, CHEESE AND BUTTER: A Practical Handbook on their Properties and the Processes of their Production, including a Chapter on Cream and the Methods of its Separation from Milk. By JOHN OLIVER, late Principal of the Western Dairy Institute, Berkeley. With Coloured Plates and 200 Illusts. Crown 8vo, 7s. 6d. cloth. [*Just published.*

"An exhaustive and masterly production. It may be cordially recommended to all students and practitioners of dairy science."—*N.B. Agriculturist.*
"We strongly recommend this very comprehensive and carefully-written book to dairy-farmers and students of dairying. It is a distinct acquisition to the library of the agriculturist."—*Agricultural Gazette.*

Agricultural Facts and Figures.

NOTE-BOOK OF AGRICULTURAL FACTS AND FIGURES FOR FARMERS AND FARM STUDENTS. By PRIMROSE McCONNELL, B.Sc. Fifth Edition. Royal 32mo, roan, gilt edges, with band, 4s.

"Literally teems with information, and we can cordially recommend it to all connected with agriculture."—*North British Agriculturist.*

Small Farming.

SYSTEMATIC SMALL FARMING; *or, The Lessons of my Farm.* Being an Introduction to Modern Farm Practice for Small Farmers. By R. SCOTT BURN. With numerous Illustrations, crown 8vo, 6s. cloth.

"This is the completest book of its class we have seen, and one which every amateur farmer will read with pleasure and accept as a guide."—*Field.*

Modern Farming.

OUTLINES OF MODERN FARMING. By R. SCOTT BURN. Soils, Manures, and Crops—Farming and Farming Economy—Cattle, Sheep, and Horses—Management of Dairy, Pigs, and Poultry—Utilisation of Town-Sewage, Irrigation, &c. Sixth Edition. In One Vol., 1,250 pp., half-bound, profusely Illustrated, 12s.

"The aim of the author has been to make his work at once comprehensive and trustworthy and he has succeeded to a degree which entitles him to much credit."—*Morning Advertiser.*

Agricultural Engineering.

FARM ENGINEERING, THE COMPLETE TEXT-BOOK OF. Comprising Draining and Embanking; Irrigation and Water Supply; Farm Roads, Fences, and Gates; Farm Buildings; Barn Implements and Machines; Field Implements and Machines; Agricultural Surveying, &c. By Prof. JOHN SCOTT. 1,150 pages, half-bound, with over 600 Illustrations, 12s.

"Written with great care, as well as with knowledge and ability. The author has done his work well; we have found him a very trustworthy guide wherever we have tested his statements. The volume will be of great value to agricultural students."—*Mark Lane Express.*

Agricultural Text-Book.

THE FIELDS OF GREAT BRITAIN: A Text-Book of Agriculture, adapted to the Syllabus of the Science and Art Department. For Elementary and Advanced Students. By HUGH CLEMENTS (Board of Trade). Second Edition, Revised, with Additions. 18mo, 2s. 6d. cloth.

"A most comprehensive volume, giving a mass of information."—*Agricultural Economist.*
"It is a long time since we have seen a book which has pleased us more, or which contains such a vast and useful fund of knowledge."—*Educational Times.*

Tables for Farmers, etc.

TABLES, MEMORANDA, AND CALCULATED RESULTS *for Farmers, Graziers, Agricultural Students, Surveyors, Land Agents, Auctioneers, etc.* With a New System of Farm Book-keeping. By SIDNEY FRANCIS. Third Edition, Revised. 272 pp., waistcoat pocket size, 1s. 6d. leather.

"Weighing less than 1 oz., and occupying no more space than a match box, it contains a mass of facts and calculations which has never before, in such handy form, been obtainable. Every operation on the farm is dealt with. The work may be taken as thoroughly accurate, the whole of the tables having been revised by Dr. Fream. We cordially recommend it."—*Bell's Weekly Messenger.*

AGRICULTURE, FARMING, GARDENING, etc. 45

Artificial Manures and Foods.
FERTILISERS AND FEEDING STUFFS: Their Properties and Uses. A Handbook for the Practical Farmer. By BERNARD DYER, D.Sc. (Lond.) With the Text of the Fertilisers and Feeding Stuffs Act of 1893, the Regulations and Forms of the Board of Agriculture and Notes on the Act by A. J. DAVID, B.A., LL.M., of the Inner Temple, Barrister-at-Law. Crown 8vo, 120 pages, 1s. cloth. [*Just published.*
"An excellent shillingsworth. Dr. Dyer has done farmers good service in placing at their disposa so much useful information in so intelligible a form."—*The Times.*

The Management of Bees.
BEES FOR PLEASURE AND PROFIT: A Guide to the Manipulation of Bees, the Production of Honey, and the General Management of the Apiary. By G. GORDON SAMSON. Crown 8vo, 1s. cloth.
"The intending bee-keeper will find exactly the kind of information required to enable him to make a successful start with his hives. The author is a thoroughly competent teacher, and his book may be commended."—*Morning Post.*

Farm and Estate Book-keeping.
BOOK-KEEPING FOR FARMERS & ESTATE OWNERS. A Practical Treatise, presenting, in Three Plans, a System adapted for all Classes of Farms. By JOHNSON M. WOODMAN, Chartered Accountant. Second Edition, Revised. Crown 8vo, 3s. 6d. cloth boards; or 2s. 6d. cloth limp.
"The volume is a capital study of a most important subject."—*Agricultural Gazette.*
"The young farmer, land agent, and surveyor will find Mr. Woodman's treatise more than repay its cost and study."—*Building News.*

Farm Account Book.
WOODMAN'S YEARLY FARM ACCOUNT BOOK. Giving a Weekly Labour Account and Diary, and showing the Income and Expenditure under each Department of Crops, Live Stock, Dairy, &c. &c. With Valuation, Profit and Loss Account, and Balance Sheet at the end of the Year. By JOHNSON M. WOODMAN, Chartered Accountant, Author of "Book-keeping for Farmers." Folio, 7s. 6d. half bound. [*culture*
"Contains every requisite form for keeping farm accounts readily and accurately."—*Agri-*

Early Fruits, Flowers, and Vegetables.
THE FORCING GARDEN; or, How to Grow Early Fruits, Flowers, and Vegetables. With Plans and Estimates for Building Glass-houses, Pits, and Frames. By SAMUEL WOOD. Crown 8vo, 3s. 6d. cloth.
"A good book, and fairly fills a place that was in some degree vacant. The book is written with great care, and contains a great deal of valuable teaching."—*Gardeners' Magazine.*

Good Gardening.
A PLAIN GUIDE TO GOOD GARDENING; or, How to Grow Vegetables, Fruits, and Flowers. By S. WOOD. Fourth Edition, with considerable Additions, &c., and numerous Illustrations. Crown 8vo, 3s. 6d. cl.
"A very good book, and one to be highly recommended as a practical guide. The practical directions are excellent."—*Athenæum.*
"May be recommended to young gardeners, cottagers, and specially to amateurs, for the plain, simple, and trustworthy information it gives on common matters too often neglected."—*Gardeners' Chronicle.*

Gainful Gardening.
MULTUM-IN-PARVO GARDENING; or, How to make One Acre of Land produce £620 a-year by the Cultivation of Fruits and Vegetables; also, How to Grow Flowers in Three Glass Houses, so as to realise £176 per annum clear Profit. By SAMUEL WOOD, Author of "Good Gardening," &c. Fifth and Cheaper Edition, Revised, with Additions. Crown 8vo, 1s. sewed.
"We are bound to recommend it as not only suited to the case of the amateur and gentleman's gardener, but to the market grower."—*Gardeners' Magazine.*

Gardening for Ladies.
THE LADIES' MULTUM-IN-PARVO FLOWER GARDEN, and Amateurs' Complete Guide. With Illusts. By S. WOOD. Cr. 8vo, 3s. 6d. cl.

Receipts for Gardeners.
GARDEN RECEIPTS. Edited by CHARLES W. QUIN. 12mo, 1s. 6d. cloth limp.

Market Gardening.
MARKET AND KITCHEN GARDENING. By Contributors to "The Garden." Compiled by C. W. SHAW, late Editor of "Gardening Illustrated." 12mo 3s. 6d. cloth boards.

AUCTIONEERING, VALUING, LAND SURVEYING ESTATE AGENCY, etc.

Auctioneer's Assistant.
THE APPRAISER, AUCTIONEER, BROKER, HOUSE AND ESTATE AGENT AND VALUER'S POCKET ASSISTANT, for the Valuation for Purchase, Sale, or Renewal of Leases, Annuities and Reversions, and of property generally; with Prices for Inventories, &c. By JOHN WHEELER, Valuer, &c. Sixth Edition, Re-written and greatly extended by C. NORRIS, Surveyor, Valuer, &c. Royal 32mo, 5s. cloth.

"A neat and concise book of reference, containing an admirable and clearly-arranged list of prices for inventories, and a very practical guide to determine the value of furniture, &c."—*Standard.*

"Contains a large quantity of varied and useful information as to the valuation for purchase, sale, or renewal of leases, annuities and reversions, and of property generally, with prices for inventories, and a guide to determine the value of interior fittings and other effects."—*Builder.*

Auctioneering.
AUCTIONEERS: THEIR DUTIES AND LIABILITIES. A Manual of Instruction and Counsel for the Young Auctioneer. By ROBERT SQUIBBS, Auctioneer. Second Edition, Revised and partly Re-written. Demy 8vo, 12s. 6d. cloth.

*** OPINIONS OF THE PRESS.

"The standard text-book on the topics of which it treats."—*Athenæum.*

"The work is one of general excellent character, and gives much information in a compendious and satisfactory form."—*Builder.*

"May be recommended as giving a great deal of information on the law relating to auctioneers, in a very readable form."—*Law Journal.*

"Auctioneers may be congratulated on having so pleasing a writer to minister to their special needs."—*Solicitors' Journal.*

"Every auctioneer ought to possess a copy of this excellent work."—*Ironmonger.*

"Of great value to the profession. . . . We readily welcome this book from the fact that it treats the subject in a manner somewhat new to the profession."—*Estates Gazette.*

Inwood's Estate Tables.
TABLES FOR THE PURCHASING OF ESTATES, Freehold, Copyhold, or Leasehold; Annuities, Advowsons, etc., and for the Renewing of Leases held under Cathedral Churches, Colleges, or other Corporate bodies for Terms of Years certain, and for Lives; also for Valuing Reversionary Estates, Deferred Annuities, Next Presentations, &c.; together with SMART'S Five Tables of Compound Interest, and an Extension of the same to Lower and Intermediate Rates. By W. INWOOD. 24th Edition, with considerable Additions, and new and valuable Tables of Logarithms for the more Difficult Computations of the Interest of Money, Discount, Annuities, &c., by M. FEDOR THOMAN, of the Société Crédit Mobilier of Paris. Crown 8vo, 8s. cloth.

"Those interested in the purchase and sale of estates, and in the adjustment of compensation cases, as well as in transactions in annuities, life insurances, &c., will find the present edition of eminent service."—*Engineering.*

"'Inwood's Tables' still maintain a most enviable reputation. The new issue has been enriched by arge additional contributions by M. Fedor Thoman, whose carefully arranged Tables cannot fail to be of the utmost utility."—*Mining Journal.*

Agricultural Valuer's Assistant.
THE AGRICULTURAL VALUER'S ASSISTANT. A Practical Handbook on the Valuation of Landed Estates; including Rules and Data for Measuring and Estimating the Contents, Weights, and Values of Agricultural Produce and Timber, and the Values of Feeding Stuffs, Manures, and Labour; with Forms of Tenant-Right-Valuations, Lists of Local Agricultural Customs, Scales of Compensation under the Agricultural Holdings Act, &c. &c. By TOM BRIGHT, Agricultural Surveyor. Second Edition, much Enlarged. Crown 8vo, 5s. cloth.

"Full of tables and examples in connection with the valuation of tenant-right, estates, labour, contents, and weights of timber, and farm produce of all kinds."—*Agricultural Gazette.*

"An eminently practical handbook, full of practical tables and data of undoubted interest and value to surveyors and auctioneers in preparing valuations of all kinds."—*Farmer.*

Plantations and Underwoods.
POLE PLANTATIONS AND UNDERWOODS: A Practical Handbook on Estimating the Cost of Forming, Renovating, Improving, and Grubbing Plantations and Underwoods, their Valuation for Purposes of Transfer, Rental, Sale, or Assessment. By TOM BRIGHT, Author of "The Agricultural Valuer s Assistant," &c. Crown 8vo, 3s. 6d. cloth.

"To valuers, foresters and agents it will be a welcome aid."—*North British Agriculturist.*

"We ll calculated to assist the valuer in the discharge of his duties, and of undoubted interest and use both to surveyors and auctioneers in preparing valuations of all kinds."—*Kent Herald.*

AUCTIONEERING, VALUING, LAND SURVEYING, etc. 47

Hudson's Land Valuer's Pocket-Book.
THE LAND VALUER'S BEST ASSISTANT: Being Tables on a very much Improved Plan, for Calculating the Value of Estates. With Tables for reducing Scotch, Irish, and Provincial Customary Acres to Statute Measure, &c. By R. HUDSON, C.E. New Edition. Royal 32mo. 4s. leather.
"Of incalculable value to the country gentleman and professional man."—*Farmers' Journal.*

Ewart's Land Improver's Pocket-Book.
THE LAND IMPROVER'S POCKET-BOOK OF FORMULÆ, TABLES, and MEMORANDA *required in any Computation relating to the Permanent Improvement of Landed Property.* By JOHN EWART, Surveyor. Second Edition. Royal 32mo, 4s. leather.

Complete Agricultural Surveyor's Pocket-Book.
THE LAND VALUER'S AND LAND IMPROVER'S COMPLETE POCKET-BOOK. Being of the above Two Works bound together. Leather, with strap, 7s. 6d.

House Property.
HANDBOOK OF HOUSE PROPERTY. A Popular and Practical Guide to the Purchase, Mortgage, Tenancy, and Compulsory Sale of Houses and Land, including the Law of Dilapidations and Fixtures; with Examples of all kinds of Valuations, Useful Information on Building, and Suggestive Elucidations of Fine Art. By E. L. TARBUCK, Architect and Surveyor. Fifth Edition, Enlarged. 12mo, 5s. cloth.
"The advice is thoroughly practical."—*Law Journal.*
"For all who have dealings with house property, this is an indispensable guide."—*Decoration.*
"Carefully brought up to date, and much improved by the addition of a division on fine art. . . . A well written and thoughtful work."—*Land Agent's Record.*

LAW AND MISCELLANEOUS.

Pocket-Book for Sanitary Officials.
THE HEALTH OFFICER'S POCKET-BOOK: A Guide to Sanitary Practice and Law. For Medical Officers of Health, Sanitary Inspectors, Members of Sanitary Authorities, &c. By EDWARD F. WILLOUGHBY, M.D. (Lond.), &c., Author of "Hygiene and Public Health." Fcap. 8vo, 7s. 6d. cloth, red edges, rounded corners. [*Just published.*
"A mine of condensed information of a pertinent and useful kind on the various subjects of which it treats. The matter seems to have been carefully compiled and arranged for facility of reference, and it is well illustrated by diagrams and woodcuts. The different subjects are succinctly but fully and scientifically dealt with."—*The Lancet.*
"An excellent publication, dealing with the scientific, technical and legal matters connected with the duties of medical officers of health and sanitary inspectors. The work is replete with information."—*Local Government Journal.*

Journalism.
MODERN JOURNALISM. A Handbook of Instruction and Counsel for the Young Journalist. By JOHN B. MACKIE, Fellow of the Institute of Journalists. Crown 8vo, 2s. cloth. [*Just published.*
"This invaluable guide to journalism is a work which all aspirants to a journalistic career will read with advantage."—*Journalist.*

Private Bill Legislation and Provisional Orders.
HANDBOOK FOR THE USE OF SOLICITORS AND ENGINEERS Engaged in Promoting Private Acts of Parliament and Provisional Orders, for the Authorization of Railways, Tramways, Gas and Water Works, &c. By L. LIVINGSTON MACASSEY, of the Middle Temple, Barrister-at-Law, M.Inst.C.E. 8vo, 25s. cloth.

Law of Patents.
PATENTS FOR INVENTIONS, AND HOW TO PROCURE THEM. Compiled for the Use of Inventors, Patentees and others. By G. G. M. HARDINGHAM, Assoc.Mem.Inst.C.E., &c. Demy 8vo, 1s. 6d. cloth.

Labour Disputes.
CONCILIATION AND ARBITRATION IN LABOUR DISPUTES: A Historical Sketch and Brief Statement of the Present Position of the Question at Home and Abroad. By J. S. JEANS, Author of "England's Supremacy," &c. Crown 8vo, 200 pp., 2s. 6d. cloth. [*Just published.*
"Mr. Jeans is well qualified to write on this subject, both by his previous books and by his practical experience as an arbitrator."—*The Times.*

A Complete Epitome of the Laws of this Country.

EVERY MAN'S OWN LAWYER: A Handy-Book of the Principles of Law and Equity. With A CONCISE DICTIONARY OF LEGAL TERMS. By A Barrister. Thirty-second Edition, carefully Revised, and including New Acts of Parliament of 1894. Comprising the Local Government Act, 1894 (establishing District and Parish Councils); Finance Act, 1894 (imposing the New Death Duties); Merchant Shipping Act, 1894; Prevention of Cruelty to Children Act, 1894; Building Societies Act, 1894; Notice of Accidents Act, 1894; Sale of Goods Act, 1893; Voluntary Conveyances Act, 1893; Married Women's Property Act, 1893; Trustee Act, 1893; Fertiliser and Feeding Stuffs Act, 1893; Betting and Loans (Infants) Act, 1892; Shop Hours Act, 1892; Small Holdings Act, 1892; and many other important new Acts. Crown 8vo, 750 pp., price 6s. 8d. (saved at every consultation!), strongly bound in cloth. [Just published.

*** The Book will be found to comprise (amongst other matter)—
THE RIGHTS AND WRONGS OF INDIVIDUALS—LANDLORD AND TENANT—VENDORS AND PURCHASERS—LEASES AND MORTGAGES—PRINCIPAL AND AGENT—PARTNERSHIP AND COMPANIES—MASTERS, SERVANTS, AND WORKMEN—CONTRACTS AND AGREEMENTS—BORROWERS, LENDERS, AND SURETIES—SALE AND PURCHASE OF GOODS—CHEQUES, BILLS, AND NOTES—BILLS OF SALE—BANKRUPTCY—RAILWAY AND SHIPPING LAW—LIFE, FIRE, AND MARINE INSURANCE—ACCIDENT AND FIDELITY INSURANCE—CRIMINAL LAW—PARLIAMENTARY ELECTIONS—COUNTY COUNCILS—DISTRICT COUNCILS—PARISH COUNCILS—MUNICIPAL CORPORATIONS—LIBEL AND SLANDER—PUBLIC HEALTH AND NUISANCES—COPYRIGHT, PATENTS, TRADE MARKS—HUSBAND AND WIFE—DIVORCE—INFANCY—CUSTODY OF CHILDREN—TRUSTEES AND EXECUTORS—CLERGY, CHURCHWARDENS, ETC.—GAME LAWS AND SPORTING—INNKEEPERS—HORSES AND DOGS—TAXES AND DEATH DUTIES—FORMS OF AGREEMENTS, WILLS, CODICILS, NOTICES, ETC.

☞ *The object of this work is to enable those who consult it to help themselves to the law; and thereby to dispense, as far as possible, with professional assistance and advice. There are many wrongs and grievances which persons submit to from time to time through not knowing how or where to apply for redress; and many persons have as great a dread of a lawyer's office as of a lion's den. With this book at hand it is believed that many a SIX-AND-EIGHTPENCE may be saved; many a wrong redressed; many a right reclaimed; many a law suit avoided; and many an evil abated. The work has established itself as the standard legal adviser of all classes, and has also made a reputation for itself as a useful book of reference for lawyers residing at a distance from law libraries, who are glad to have at hand a work embodying recent decisions and enactments.*

*** OPINIONS OF THE PRESS.

"A complete code of English Law, written in plain language, which all can understand. ould be in the hands of every business man, and all who wish to abolish lawyers' bills. . . ."—*Weekly Times.*

"A useful and concise epitome of the law, compiled with considerable care."—*Law Magazine.*

"A complete digest of the most useful facts which constitute English law."—*Globe.*

"Admirably done, admirably arranged, and admirably cheap."—*Leeds Mercury.*

"A concise, cheap and complete epitome of the English law. So plainly written that he who runs may read, and he who reads may understand."—*Figaro.*

"The latest edition of this popular book ought to be in every business establishment, and on every library table."—*Sheffield Post.*

"A complete epitome of the law; thoroughly intelligible to non-professional readers."—*Bell's Life.*

Legal Guide for Pawnbrokers.

THE LAW OF LOANS AND PLEDGES. With Statutes and a Digest of Cases. By H. C. FOLKARD, Esq., Barrister-at-Law. Fcap. 8vo, 3s. 6d. cloth.

The Law of Contracts.

LABOUR CONTRACTS: A Popular Handbook on the Law of Contracts for Works and Services. By DAVID GIBBONS. Fourth Edition, Appendix of Statutes by T. F. UTTLEY, Solicitor. Fcap. 8vo, 3s. 6d. cloth.

The Factory Acts.

SUMMARY OF THE FACTORY AND WORKSHOP ACTS (1878-1891). For the Use of Manufacturers and Managers. By EMILE GARCKE and J. M. FELLS. (Reprinted from "FACTORY ACCOUNTS.") Crown 8vo, 6d. sewed.

www.ingramcontent.com/pod-product-compliance
Lightning Source LLC
Chambersburg PA
CBHW020902230426
43666CB00008B/1276